The Joy of Geometry

Alfred S. Posamentier

Prometheus Books

Guilford, Connecticut

 Prometheus Books

An imprint of The Rowman & Littlefield Publishing Group, Inc.
4501 Forbes Blvd., Ste. 200
Lanham, MD 20706
www.rowman.com

Distributed by NATIONAL BOOK NETWORK

British Library Cataloguing in Publication Information available

Library of Congress Control Number Available
ISBN 9781633885868 (pbk. : alk. paper)
ISBN 9781633885875 (electronic)

♾™ The paper used in this publication meets the minimum requirements of American National Standard for Information Sciences—Permanence of Paper for Printed Library Materials, ANSI/NISO Z39.48–1992

To my children and grandchildren, whose future is unbounded,
Lisa, Daniel, David, Lauren, Max, Samuel, Jack, and Charles.

Contents

Introduction

We know that geometry is all around us. Yet many people haven't had the opportunity to appreciate amazing geometric relationships and their beauty. The high school curriculum in the United States typically designates one year for the study of geometry providing an analog of the work of a mathematician, who builds up a field of study beginning with accepted knowledge—axioms and postulates—and then progresses to proving theorems in a logical order. However, this concentration on the proof of theorems bypasses many unusual relationships. This book attempts to present these geometric wonders without the "distraction" of proofs.

The modern understanding of geometry in the English language began largely in the 18th century when the Scottish mathematician Robert Simson published an English version of a large portion of Euclid's elements. Throughout the 19th century, the information was further refined by the French mathematician Adrien-Marie Legendre (1752–1833). Geometry's first import into the United States was by the American mathematician Charles Davies (1798–1876), who wrote the standard textbook for a course that began with definitions, axioms, and postulates, which led to theorems. This was originally a college-level course, and in the 20th century it began to be introduced typically in the 10th grade of high school. Most people's recollection is of proving theorems, which eventually builds up a body of knowledge in the way a mathematician approaches a study of mathematics.

There is clearly beauty in developing a body of knowledge where one step depends on the previous steps. However, very often this does not allow for students to truly appreciate the amazing relationships that permeate the subject of plane geometry. Under totally unexpected situations, you can find three or more lines concurrent (lines containing a common point) or three or more points concurrent (points on the same straight line). Or, if you know that any three noncollinear points determine a unique circle, under what circumstances will four or more points lie on the same circle? These are just some of the surprising elements of geometry that have often gone unnoticed. I hope to rectify this oversight in this book.

Besides these unexpected characteristics, many aspects of geometry have been—perhaps inadvertently—neglected the way the subject matter has been presented. Naturally, when dealing with linear figures, the triangle dominates. However, the quadrilateral (a four-sided figure) also deserves consideration, which will be introduced at the appropriate time. To whet your appetite, consider drawing an ugly quadrilateral, where no sides are parallel or the same length. If you draw another quadrilateral by consecutively joining the midpoints of the awkward-looking quadrilateral you have drawn, you will always end up with a parallelogram (as shown in Figure 0.1). This is clearly not expected, but it is true! There is much more that can be done with this and many other such geometric novelties, as you will see in our journey through geometry.

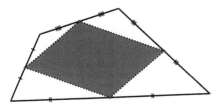

Figure 0.1. Parallelogram formed by joining the midpoints of the quadrilateral.

Much of geometry is not only a function of a lot of interesting discoveries but also the result of brilliant work by famous mathematicians over the centuries. This book will explore some of their findings in the context of the geometric relationships that evolved from their work.

There is also beauty in geometry. The golden rectangle is an example. In this book this figure will be presented in such a way that it can be appreciated both aesthetically and mathematically. Naturally, the golden rectangle permeates many other branches of mathematics, so we will touch on just a few of these to highlight its importance.

Geometry can also be entertaining, as when we look at the mistakes that can be made in it—errors that often go completely unnoticed until an absurd result evolves. Then we become concerned and look to rectify what might have been wrongly worked or assumed.

In this book, we will begin our journey through geometry to admire its wonders by considering concurrency of lines. We are interested here primarily in the relationships that exist, so the flow of the book will not be disturbed with proofs. As mentioned earlier, high school geometry courses consist largely of developing a logical system within geometry, which highlights proofs more than the results that are proved. This book, in contrast, will look at those results and admire the amazing relationships that exist. This will be the modus operandi throughout. Lots of references will be provided at the end of the book. These will enable readers to access proofs of many of the concepts that are presented without proof. The goal here is to allow you, the reader, to appreciate geometry and not be distracted by proving each of the findings presented.

As you encounter these truly unexpected and rather interesting relationships, you may be tempted to dig into your geometric toolbox and attempt to prove them. Or you might do a geometric construction with an unmarked straight edge and a pair of compasses. Today, however, we can convince ourselves that a relationship truly holds by using various dynamic geometry software products such as *Geometer's Sketchpad* (www.keycurriculum. com/) or *Geogebra*; the latter can be obtained without cost at www.geogebra. org/?lang=en. In short, demonstrating the many unexpected relationships that appear throughout the book by using one of these software products can be almost as convincing as a proper logical proof. Frankly, you should expect to regularly exclaim, "WOW, what an amazing result!" while investigating the marvels shown in this book. We invite you now to join us on this rather unusual approach to appreciating geometry's power and beauty, unencumbered by proofs.

1

Concurrent Lines

As one of the basic elements in geometry, lines deserve investigation. We know that any two nonparallel lines eventually will intersect. But when a third line shares a point of intersection with two other lines, we have three *concurrent lines*, which share a common point. This relationship becomes more interesting when more than three lines share a common point. We begin with various concurrent lines that are a part of the basic triangle relationships. Then, after we establish these rather common concurrencies, we will extend our knowledge of concurrency to other geometric figures. At that point, we expect to provide the reader with ideas that should elicit surprise and amazement. Let's begin by considering the three altitudes of a triangle, lines from the vertices of a triangle that are perpendicular to the opposite sides.

THE ALTITUDES OF A TRIANGLE

Perhaps the most common concurrency is that generated by the three altitudes of a triangle. Generally, we take this fact for granted. However, it is a good example with which to begin our consideration of concurrent lines. In Figures 1.1, 1.2, and 1.3 we show the three basic types of triangles: an acute triangle, where no angle exceeds 90°; a right triangle; and an obtuse triangle, where one angle is greater than 90°. Each of these triangles also has the three altitudes AD, BE, and CF intersecting at point H. This point of intersection, often referred to as the *point of concurrency*, is called the *orthocenter* of the triangle and is located inside an acute triangle, outside an obtuse triangle, and at the right-angle vertex of a right triangle, as shown in Figures 1.1 through 1.3.

1

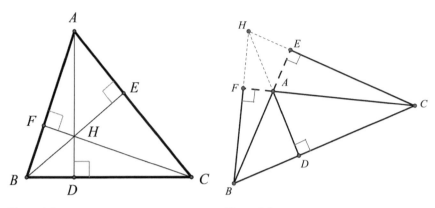

Figure 1.1.   **Figure 1.2.**

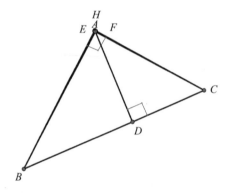

Figure 1.3.

We can gather much more information from this altitude relationship. The point of concurrency is particularly interesting in that it divides the altitudes into lengths that yield equal products. Although this relationship holds true for all three situations pictured in Figures 1.1, 1.2, and 1.3, it is probably easiest to see for the acute triangle, where $AH \cdot HD = BH \cdot HE = CH \cdot HF$.

Unfortunately, this relationship is rarely shown to high school classes, although it easily could be presented when discussing similar triangles. Also not shared in high school geometry is another relationship that further enhances knowledge about the three altitudes of a triangle. The triangles shown in Figures 1.1, 1.2, and 1.3 demonstrate a relationship involving the alternate segments determined by the feet of the altitudes (the points at which the altitudes intersect the opposite sides): $BD^2 + CE^2 + AF^2 = CD^2 + AE^2 + BF^2$.

Altitudes do not stand alone in their relationship to other triangle parts. For example, in Figure 1.4, we begin with a triangle and one of its altitudes, say, *AE*. Then we draw the radius of the circumcircle (the circle containing the three vertices of the triangle) to vertex *A*. Unexpectedly, when we draw the angle bisector of angle *A* of triangle *ABC*, we find that this bisector also bisects the angle we just created, angle *IAE*; or, put another way, $\angle EAD = \angle IAD$. This relationship, one of many that unfortunately are not shown to high school classes, foreshadows the well-hidden relationships that we will examine going forward.

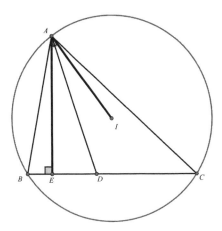

Figure 1.4.

Another example of how an altitude of a triangle can lead to an angle bisector is shown in Figure 1.5, where we see triangle *ABC* with altitude *AD*. Two random lines are drawn from points *B* and *C* that intersect at a point *E* on the altitude and meet sides *AC* and *AB* of the triangle at points *P* and *Q*, respectively. This makes the altitude *AD* bisect angle *PDQ*, so that $\angle ADQ = \angle ADP$. This example demonstrates how the altitude can relate to an angle bisector—which in this case is the altitude itself!

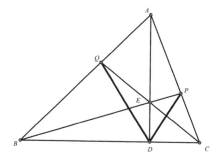

Figure 1.5.

CONSIDERING OTHER CONCURRENCIES

Concurrencies sometimes occur in the strangest ways. Consider, for example, triangle *ABC*, shown in Figure 1.6. Here, side *BC* is extended to point *P*, which can be placed anywhere along the extended side. We then draw a random line from *P* to intersect sides *AC* and *AB* of triangle *ABC* at points *D* and *F*, respectively. When we draw *DE* parallel to *AB*, and *FE* parallel to *AC*, lines *EF*, *DE*, and *BC* all contain point *E*, or are concurrent at point *E*. Recall that we could select point *P* anywhere on the extension of *BC*, which is what makes this example so unusual.

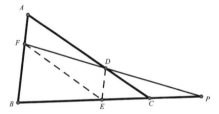

Figure 1.6.

Many other relationships involve the altitudes of a triangle. For example, if any altitude is extended to the circumcircle of the original triangle, the side of the triangle (*BC*) bisects the line segment from the orthocenter to the point of intersection with the circumcircle. In Figure 1.7, referring to altitude *AD*, note that point *D* is the midpoint of *HG*, or put another way, $HD = GD$.

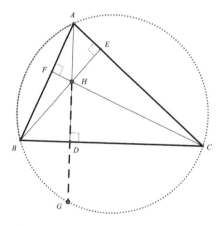

Figure 1.7.

If we extend another altitude, as shown in Figure 1.8, we now have two altitudes that meet the circumcircle at points *G* and *J*. Rather unexpectedly, this determines two equal arcs, *JC* and *GC*, on the circle; or, put another way, point *C* bisects arc *JCG*. This is true for any inscribed triangle that has two altitudes protruding toward the circumcircle. That's what makes this relationship so interesting.

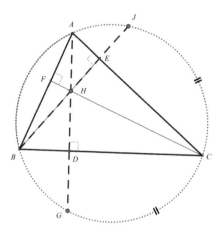

Figure 1.8.

Suppose we now extend the remaining altitude, *CF*, to intersect the circle at point *K*. When we connect the three points of contact of the altitudes, *J, K*, and *G*, with the circumcircle, the result is not only a triangle similar to the one formed by the feet of the altitudes (called an *orthic triangle*) but also a situation in which the corresponding sides of these two similar triangles are parallel. This is shown in Figure 1.9.

Particularly surprising in Figure 1.9, since this is an acute triangle, is that the altitudes of triangle *ABC* bisect the angles of the orthic triangle. In other words, altitude *AD* bisects angle *EDF*, altitude *DE* bisects angle *FED*, and altitude *CF* bisects angle *DFE*.

The positioning of the orthic triangle is also intriguing. When radii of the circumcircle of a triangle are drawn to a vertex of the triangle, they are perpendicular to each of the sides of the orthic triangle. This is shown in Figure 1.10 where for triangle *ABC*, the radii of circumcircle *O* are drawn to each of the three vertices, *A, B*, and *C*, and the radii are perpendicular to the sides of the orthic triangle, *DEF*. Such simple yet unexpected relationships go far to enhance the beauty of geometry.

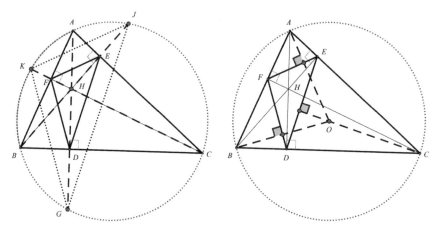

Figure 1.9.   **Figure 1.10.**

A triangle is said to be inscribed in a second triangle, if each of its vertices is on a side of the larger triangle. In Figure 1.11, triangle *DEF* is inscribed in triangle *ABC*. For this acute triangle *ABC*, however, we have constructed triangle *DEF* so that each of its vertices is on the foot of one of the larger triangle's altitudes. We can inscribe a second triangle in triangle *ABC*, as shown in Figure 1.11. The smallest perimeter of all the possible inscribed triangles of the original acute triangle *ABC* is that of triangle *DEF* formed with its vertices at the feet of the altitudes. That is, the perimeter of triangle *DEF* is less than that of triangle *XYZ* or any other triangle formed by three points on the sides of the original triangle. Recall from an earlier example that each of the altitudes of the original triangle *ABC*, namely *AD*, *DE*, and *CF*, bisects an angle of the orthic triangle.

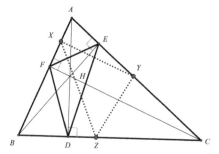

Figure 1.11.

A slight digression here might be of interest. For any point within an equilateral triangle, such as point P in triangle ABC (Figure 1.12), we find that the sum of the perpendicular distances from P to each of the three sides of the equilateral triangle is the same as that for any other point, say Q, selected within the triangle. The distance sum is always equal to the length of the altitude of the equilateral triangle. Therefore, referring to Figure 1.12, we can summarize this relationship as follows: $PK + PH + PD = QJ + QG + QF = AE$.

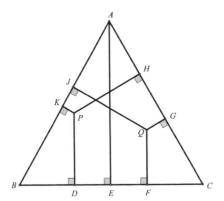

Figure 1.12.

A peculiar relationship evolves when a semicircle is drawn on one side of an equilateral triangle, as shown in Figure 1.13. Here we have triangle ABC with semicircle $ABDC$ drawn on side AC. Points E and F are the trisection points of line segment BC. We then draw line AE to meet the semicircle at G, and line AF to meet the semicircle at J. Quite surprisingly, the semicircle is also trisected by these three lines, so that BG, GJ, and CJ are equal arcs.

To see how inclusive the orthocenter is with the remainder of the triangle, consider the midpoint of two of the altitudes and one of the sides. This is shown in Figure 1.14, where N is the midpoint of altitude CF, M is a midpoint of altitude BE, and K is the midpoint of side BC. It is well known that any three noncollinear points determine a unique circle. However, getting other points on that circle is no mean feat. Interestingly, no matter what the shape of the original triangle ABC is, the orthocenter will also lie on the circle determined by the three noncollinear points M, N, and K, as shown in Figure 1.14. We will consider many more concyclic points (points that lie on a common circle) later on.

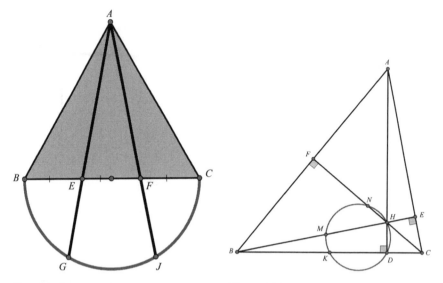

Figure 1.13.   **Figure 1.14.**

The point of intersection of the altitudes, or the orthocenter, of a triangle has many special properties. One of these is rather curious. We know that any three noncollinear points determine a unique circle. Figure 1.15 shows a circle containing the orthocenter and two vertices of the triangle. This circle, containing points B, H, and C, turns out to be congruent to the circumcircle, containing points A, B, and C, of the triangle. This same relationship can be seen for the other two circles (shown) that contain the orthocenter and a different pair of vertices of triangle ABC; that is, all four circles shown in this figure are equal in size.

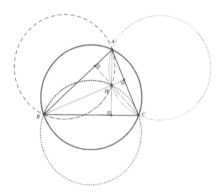

Figure 1.15.

Many other relationships can be found regarding the altitudes of a triangle and its circumcircle. One is that the distance from the circumcenter to a side of the triangle is half the distance from the orthocenter to the opposite vertex. We see this in Figure 1.16, where the distance between the circumcenter O and the side AC is measured by the perpendicular distance OG. With simple measurement we find that $BH = 2OG$.

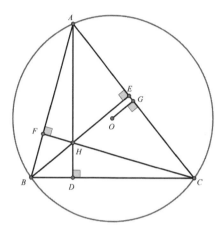

Figure 1.16.

THE CONCURRENCY OF THE MEDIANS OF A TRIANGLE

We have seen how the altitudes of a triangle are divided by their point of concurrency, where the product of the segments of each altitude is the same for all the altitudes. The medians of a triangle also are concurrent at a point, yet this point trisects each of the medians. In Figure 1.17 the three medians AD, BE, and CF meet at point G, their point of concurrency, which trisects each of the medians, so that $GD = \dfrac{1}{3}AD$, $GE = \dfrac{1}{3}BE$, and $GF = \dfrac{1}{3}CE$. This point of concurrency is called the *centroid* of the triangle, since it is the triangle's center of gravity. This means that if you wish to balance a cardboard triangle, the point at which that triangle will balance perfectly is the centroid.

Furthermore, with the medians drawn as in Figure 1.17, the triangle ABC is then partitioned into six smaller triangles of equal area.

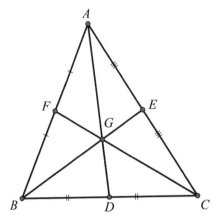

Figure 1.17.

Concurrency of lines sometimes appears when we least expect it. One such case is a line parallel to one side of a triangle, at any distance away from that side. The points of intersection of the lines joining two vertices of the triangle to the side-intersection points of the parallel line will always be concurrent with the median from a third vertex. This is shown in Figure 1.18, where *PQ* is parallel to side *BC* of triangle *ABC*. Quite surprisingly, lines *PC* and *QB* will always be concurrent with the median *AM*.

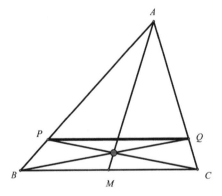

Figure 1.18.

Were we to pass line *PQ*, which is parallel to side *BC* of triangle *ABC*, through the midpoint, *N*, of *AM*, we would find that *AM* bisects *PQ*; that is, the intersection, *N*, of *AM* and *PQ* is the midpoint of segment *PQ*, which we can see in Figure 1.19. Notice that when two lines, such as *AM* and *PQ*, bisect each other, they form the diagonals of a parallelogram, *AQMP*.

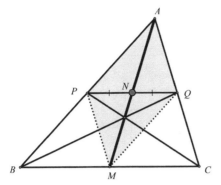

Figure 1.19.

CEVA'S THEOREM

One of the most useful, yet often neglected, theorems involving concurrency of lines of a triangle was published in 1678 by the Italian mathematician Giovanni Ceva (1648–1734) in his work *De lineis rectis.* An earlier proof of this theorem had been done by the Arab mathematician Al-Mu'taman ibn Hūd in the eleventh century. Nevertheless, we still credit it to Ceva, as he is believed to have developed it without any knowledge of its previous discovery. The theorem states that the three line segments connecting the vertices of a triangle to the opposite sides are concurrent, if and only if, the products of the alternate segments along the sides are equal. In Figure 1.20, $AF \cdot BD \cdot CE = FB \cdot DC \cdot EA$, if and only if, the three line segments AD, BE, and CF are concurrent. Using Ceva's theorem, it is trivial to prove that the medians of a triangle are concurrent, since the two products of alternate segments are clearly identical. Ceva's theorem is extremely useful in establishing concurrency of lines joining vertices to the opposite sides of a triangle, as we shall see going forward.

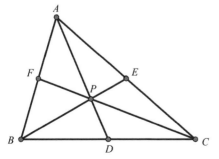

Figure 1.20.

THE CONCURRENCY OF THE ANGLE BISECTORS
OF A TRIANGLE

Yet another concurrency that should be introduced in high school geometry is that of the angle bisectors of a triangle. Figure 1.21 shows the concurrency of the angle bisectors, *AD*, *BE*, and *CF*, meeting at their point of concurrency, *I*. This point is called the *incenter*, as it is the center of the circle inscribed in the triangle, which is the circle tangent to each of the three sides of the triangle.

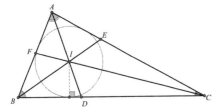

Figure 1.21.

Ceva's theorem is quite useful for proving concurrency when applied to the three angle bisectors of a triangle. Ceva's theorem would also prove that the interior angle bisector of a triangle is concurrent with the two exterior angle bisectors of the other two angles, as shown in Figure 1.22. Here the interior angle bisector *AL*, when extended, meets the two exterior angle bisectors, *KB* and *NC*, at point *P*. The ambitious reader may want to prove this concurrency using Ceva's theorem. However, here we merely appreciate the fact that fascinating relationships such as this one exist in plane geometry.

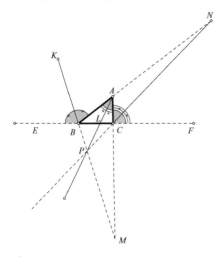

Figure 1.22.

The angle bisector of a triangle yields many unusual concurrencies. We will show one here as an example, but others will be revealed going forward. In Figure 1.23, *AD* is the bisector of angle *BAC*, and points *M* and *N* are the points of tangency of the inscribed circle to sides *AC* and *BC* of the triangle. The intersection of *MN* and *AD* is point *P*, and interestingly enough, when we draw the perpendicular line from point *B* to *AD*, it falls precisely at point *P*.

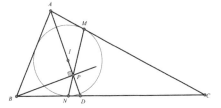

Figure 1.23.

Another example of a concurrence using both the inscribed and the circumscribed circles of a triangle is shown in Figure 1.24. Here, we have triangle *ABC*, and the bisector of angle *BAC* is line *AD*. At point *D* we erect a perpendicular to side *BC*, which meets the diameter of the circumcircle at point *P*. Quite unexpectedly, when we draw the perpendicular bisector of *AD*, it is concurrent with the other two lines at point *P*.

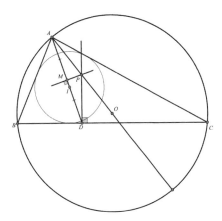

Figure 1.24.

DISCOVERING CONCURRENCIES IN TRIANGLES

Here, we show how geometric relationships can evolve from a group of midpoints in a general triangle. Consider triangle *ABC* (Figure 1.25) with three concurrent lines, *AL*, *BM*, and *CN*, drawn from each of the vertices to each of the sides, where the midpoints of each of these three lines are *P*, *Q*, and *R*, respectively. The midpoints of the three sides of the triangle are *D*, *E*, and *F*.

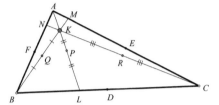

Figure 1.25.

We then create triangle *DEF* (Figure 1.26). Its sides are parallel to those of the original triangle *ABC*, since, when you join the midpoints of two sides of a triangle, the line segment formed is half the length of, and parallel to, the third side. But now we return to our search for another set of three concurrent lines.

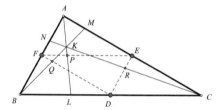

Figure 1.26.

Suppose that we join the midpoints of the sides of the original triangle *ABC* with the three midpoints of *AL*, *BM*, and *CN*, namely, *P*, *Q*, and *R*, respectively. We, unexpectedly, get another set of concurrencies, namely, lines *PD*, *QE*, and *RF*, which all contain point *S*, as shown in Figure 1.27.

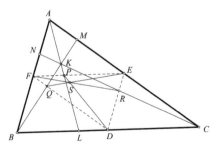

Figure 1.27.

Concurrencies can generate other concurrencies, as we have just seen. Yet, there is probably no limit to finding other such relationships. We simply select any point *P* in triangle *ABC* and connect it to each of the vertices, as shown in Figure 1.28. This gives us three concurrent line segments, *AP*, *BP*, and *CP*. We now draw the angle bisectors of each of the angles at vertex *P*. This gives us *PF* as the bisector of angle *APB*, *PE* as the bisector of angle *APC*, and *PD* as the bisector of angle *BPC*, where points *D*, *E*, and *F* are each on the sides of triangle *ABC*. When we draw lines *AD*, *BE*, and *CF*, we find that they are concurrent.

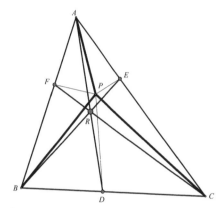

Figure 1.28.

We can also randomly select some concurrent lines in a triangle, such as those shown in Figure 1.29, where a random point *P* is chosen inside triangle *ABC*. We then locate the midpoints *M*, *K*, and *N* of the sides of triangle *ABC*, namely, *AB*, *BC*, and *CA*, respectively. When we draw lines *KL*, *NJ*, and *MG* parallel to *AP*, *BP*, and *CP* through points *K*, *N*, and *M*, respectively, we find that they are concurrent at point *Q*. Remember that point *P* was randomly chosen, so the lines are concurrent at point *Q* regardless of where point *P* is placed in triangle *ABC*.

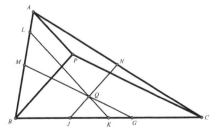

Figure 1.29.

Sometimes a concurrency can generate a second concurrency. Here we show a rather unusual arrangement of related concurrencies. We begin with triangle *ABC*, as shown in Figure 1.30. We then draw any three concurrent lines in the triangle, as we have done here with *AP*, *BP*, and *CP*. We now construct triangle *DEF* in such a way that each side of the triangle is perpendicular to one of the concurrent lines of the original triangle *ABC*. Figure 1.30 shows that *DE* is perpendicular to *AP*, *DF* is perpendicular to *BP*, and *EF* is perpendicular to *CP*. We then draw lines from the vertices of triangle *DEF* in such a way that each is perpendicular to one of the sides of triangle *ABC*. Here we see that *DK* is perpendicular to *AB*, *EL* is perpendicular to *AC*, and *FM* is perpendicular to *BC*. Unexpectedly, when these three lines are extended, we find that *DQ*, *EQ*, and *FQ* are concurrent at point *Q*. This concurrency, although it took us a bit of time to find, further demonstrates the kind of unusual patterns discoverable in geometry.

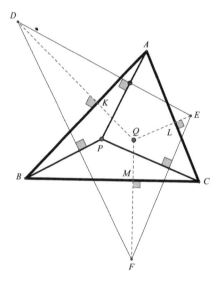

Figure 1.30.

Sometimes we can create a point of concurrency that has further surprising significance. Consider triangle *ABC* in Figure 1.31, where we trisect *BC* at points *D* and *E*. This enables us to trisect the triangle itself, so that triangles *ABD*, *ADE*, and *AEC* all have the same area. This is easy to see, since the bases are equal and the altitude from *A* to line *BC* is the same for all three triangles. We would like to find a concurrency that would give us another way to partition this triangle into three equal areas. This can be done as follows: We construct line *DF* to be parallel to *AB*, and line *EJ* to be parallel to *AC*. The point at which *EJ* intersects *DF*, which we call *P*, will allow us to partition the triangle into three equal areas, namely *APB*, *APC*, and *BPC*. We encountered a similar situation earlier (Figure 1.17), when we noted that the medians of a triangle partition the triangle into six equal-area triangles. Consequently, when these triangles are taken in pairs, the original triangle would be seen as partitioned into three equal-area triangles.

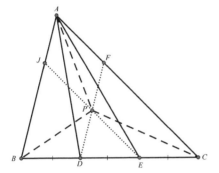

Figure 1.31.

In our examination of concurrencies of lines, we now introduce circles that are related to triangles. These include inscribed and circumscribed circles as well as circles that intersect the triangle's sides at six different points.

Let's begin by considering concurrencies that evolve from a triangle with its inscribed circle. We noted earlier that the center is determined by the concurrency of the three angle bisectors of the triangle. We show this in Figure 1.32, where the angle bisectors are *AD*, *BE*, and *CF*, which are concurrent at point *I*. Once we have inscribed the circle, we are prepared to construct another set of three concurrent lines. These lines join the triangle's vertices with the points of tangency, *T*, *U*, and *V*, of the inscribed circle with the opposite vertices, *AT*, *BU*, and *CV*, which are concurrent at point *K*. This is called the Gergonne point after its discoverer, the French mathematician Joseph-Diaz Gergonne (1771–1859).

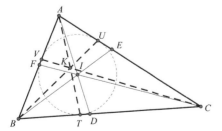

Figure 1.32.

The inscribed circle of a triangle can provide us several other surprising concurrencies. Some of these may seem contrived; however, we will consider two more concurrencies that we hope will motivate the reader to search for more such relationships.

In Figure 1.33, we draw the diameters of the inscribed circle emanating from its three points of tangency, *T*, *U*, and *V*, to meet the inscribed circle at points *M*, *N*, and *Q*, respectively. We then connect each of the triangle's vertices with these points to get *AM*, *BN*, and *CQ*. When these latter lines are extended, they are rather unexpectedly concurrent at point *P*. What makes this relationship special is that it is not well-known and it is applicable to all triangles.

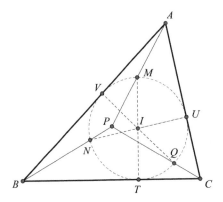

Figure 1.33.

Suppose we now do an analogous construction. We first draw *any* three concurrent lines (at point *R*). This time, however, they do not contain the center of the inscribed circle, and they emanate from the three points of tangency (*T*, *U*, and *V*) and intersect the other side of the circle at points *W*, *Y*, and *Z*, as shown in Figure 1.34. Connecting each of the newly established points *W*, *Y*, and *Z* to the nearest vertices, we find, once again, another concurrency; namely, that *AW*, *BY*, and *CZ*, when extended, meet at point *P*.

Sometimes, what appears to be a somewhat convoluted diagram can lead to a quite unexpected concurrency. Consider triangle *ABC*, shown in Figure 1.35, where the altitudes *AY*, *BZ*, and *CX* meet at point *Q*, the orthocenter. We can easily locate the center of the circumscribed circle by taking the perpendicular bisectors of any two sides of the triangle, which

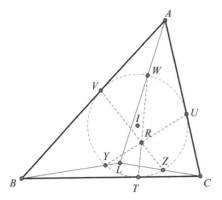

Figure 1.34.

gives us point *P*. We then locate the midpoints of *AP*, *BP*, and *CP* as points *G*, *H*, and *K*. Unexpectedly, we find that line segments *GD*, *HE*, and *FK* are concurrent at point *O*.

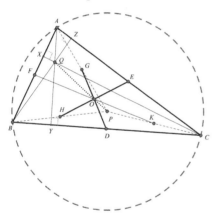

Figure 1.35.

Note also, the orthocenter in this configuration, since it gives us an added wonder, namely, the collinearity of points *Q*, *O*, and *P*. Therefore, this figure shows not only a concurrency but also a collinearity. This foreshadows the next chapter.

In our search for another concurrency, we first locate in Figure 1.36 the midpoints of the segments that connect the orthocenter with each of the vertices; that is, the midpoints of the segments *AQ*, *BQ*, and *CQ* are *G*, *H*, and *K*. We then connect these midpoints with those of the three sides of the triangle, namely *D*, *E*, and *F*. Once again, we find unexpectedly that *GD*, *HE*, and *FK* are concurrent and, in addition, they are the same length! This is quite an amazing relationship that goes completely unnoticed in high school geometry.

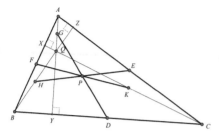

Figure 1.36.

There is no limit to the various concurrencies that can be found within a triangle. For this example, we consider two triangles that share the same inscribed circle. These are shown in Figure 1.37, where triangles *ABC* and *DEF* share the inscribed circle with center *I*.

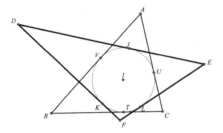

Figure 1.37.

However, there is one additional stipulation. The triangles are so situated that the lines joining their opposing vertices, *AF*, *BE*, and *DC*, are concurrent at point *P*, as shown in Figure 1.38.

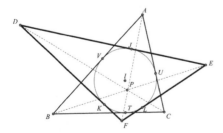

Figure 1.38.

Unexpectedly, when this happens, the lines *TJ*, *UK*, and *VL* joining their opposing points of tangency are also concurrent, as shown in Figure 1.39. One more unexpected feature is that these additional lines are concurrent at the *very same point*, point *P*, as the previous three lines. This is truly unusual!

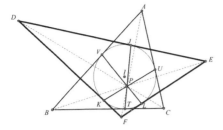

Figure 1.39.

Sometimes, segment lengths can determine concurrency. One such example is triangle ABC, shown in Figure 1.40, where point P is placed on side BC so that $AB + BP = AC + CP$. Similarly, although not clearly demonstrated in the diagram, point Q is placed on side AC so that $BC + CQ = AB + AQ$. Furthermore, point R is placed on side AB so that $BC + BR = AC + AR$. When all of these conditions are properly met, we find, curiously enough, that AP, BQ, and CR are concurrent at point P.

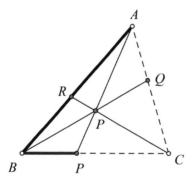

Figure 1.40.

It is always intriguing when one concurrency leads unexpectedly to another that seems completely unrelated. We see within triangle ABC (Figure 1.41) three randomly drawn lines, AL, BM, and CN, concurrent at point P. They allow us to draw the resulting triangle, MNL. We then locate the midpoints, S, Q, and R, of the sides of triangle MNL, MN, ML, and NL, respectively. Quite surprisingly, lines AS, BR, and CQ (each extended) also turn out to be concurrent, at point X.

There is an interesting variation of the previous example. Instead of using the midpoints of the sides of triangle MNL, we simply select other points, S, R, and Q, on the sides of triangle MNL so that LS, MR, and NQ are concurrent at

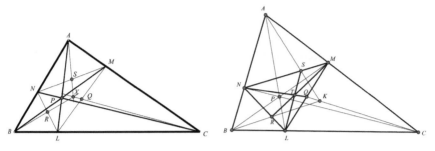

Figure 1.41.   **Figure 1.42.**

point *T*, as shown in Figure 1.42. This makes lines *AS*, *BR*, and *CQ* concurrent as well at point *K*. Remember, this depends on triangle *MNL* maintaining the concurrency at point *P*.

CIRCLES AND TRIANGLES

Our next relationship provides us some truly amazing geometry. Consider a circle intersecting a random triangle at six points, as shown in Figure 1.43. This is not just a random circle intersecting the triangle at six points, however, but rather one where lines *AD*, *BF*, and *CE* are concurrent at point *P*. (When you try to construct this, begin with the three concurrent lines and then construct a circle containing the three points of intersection with the sides.)

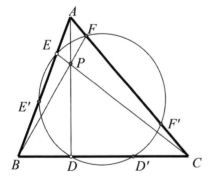

Figure 1.43.

When this happens, amazingly, the vertex connections to the other three points of intersection of triangle *ABC* and the circle, *D′*, *E′* and *F′*, determine another three lines, *AD′*, *BF′* and *CE′*, which turn out to be concurrent at point *R*, as shown in Figure 1.44.

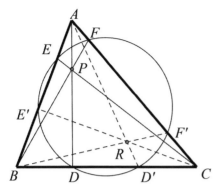

Figure 1.44.

Analogously, we now select a point P in triangle ABC, as shown in Figure 1.45. From point P we draw perpendiculars, PD, PE, and PF, to each of the three sides of the triangle. We know that any three noncollinear points determine a unique circle, so we then draw the circle determined by points D, E, and F.

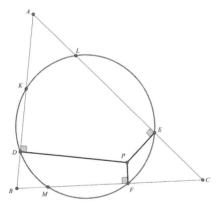

Figure 1.45.

Note that the circle also intersects the triangle at three additional points, K, L, and M, as shown in Figure 1.46. We then draw the perpendiculars to the sides of triangle ABC at each of the points K, L, and M. As you can see, these perpendiculars are concurrent. Remember, we began with a randomly selected point P, then allowed the circle formed by the perpendiculars to determine another three points that seem to be unrelated to the first three points. But, lo and behold, those new three points similarly determine a point of concurrency of perpendiculars.

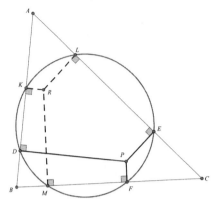

Figure 1.46.

TANGENT CIRCLES

We have introduced a circle that is neither inscribed in nor circumscribed about a triangle. Now let's do the reverse and consider a circle circumscribed about a triangle and a circle inscribed in the same triangle. This will lead us to some truly unexpected concurrencies. In Figure 1.47, the circle with center O is inscribed in the triangle and tangent to the sides at points T, U, and V. The circle with center I is circumscribed about triangle ABC, with the perpendicular bisectors of the sides (which determine the center of the circumscribed circle) meeting the circle at points K, L, and J.

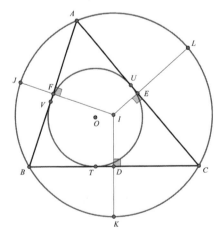

Figure 1.47.

The last determined points, K, L, and J, when joined to the points of tangency (T, U, and V) on the inscribed circle, provide us with three concurrent lines, as shown in Figure 1.48.

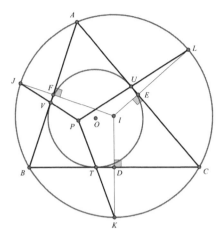

Figure 1.48.

As you might notice, points *P*, *O*, and *I* in Figure 1.49 appear to be collinear; that is, all three points lie on the same line. This, in fact, is true, as we can show. Take time to marvel over these concurrencies and collinearities, as they are often overlooked aspects of geometry that make it fascinating. We will consider collinear points in the next chapter. But since points *P*, *O*, and *I* are so obviously lined up, we have taken the liberty of mentioning them in advance.

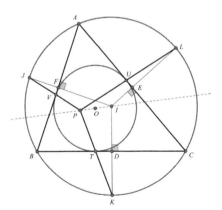

Figure 1.49.

We now extend our concept of inscribed circles beyond triangles to consider circles that are tangent to the circumscribed circle of a triangle and to a side of the triangle. Figure 1.50 shows such a configuration. The easiest way to construct this is to draw the perpendicular bisectors of the sides (which, of course, are concurrent at the center of the circumscribed circle) and then determine the midpoints between the two points of tangency. Once you have these midpoints, you have the center of the circles as well as their radii, allowing for construction of the three circles.

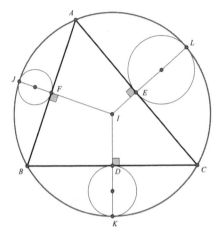

Figure 1.50.

A surprising concurrency is obtained by joining each of the common points of tangency of the pairs of circles with the opposite vertex of the triangle. This is shown in Figure 1.51, where lines *AK*, *BL*, and *CJ* are concurrent at point *P*. Remember, we began with a randomly drawn triangle, which is what makes this result so spectacular.

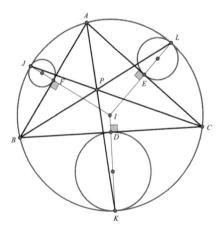

Figure 1.51.

As we approach our next unexpected concurrency, we must first recognize that when we join the midpoints of the three sides of a triangle (as shown in Figure 1.52), the triangle is divided into four congruent triangles. Furthermore, if you look at this figure carefully, you will see three parallelograms, *AEFD*, *DFEC*, and *BFED*, as well.

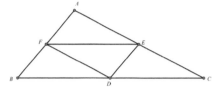

Figure 1.52.

If we now construct three inscribed circles in each of the triangles, as shown in Figure 1.53, we can establish another concurrency by connecting the centers of each of the circles with the remote vertex of the center triangle. We then have the concurrent lines *QD*, *ER*, and *FS* meeting at point *P*, another example of unexpected concurrency that demonstrates geometry's beauty.

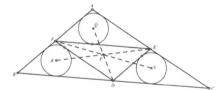

Figure 1.53.

This configuration also allows us to determine yet another concurrency by drawing lines from the vertices of the large triangle *ABC* through the center of the nearest small circle, as shown in Figure 1.54. Here you can see that lines *AQ*, *BP*, and *CS* meet at point *P*.

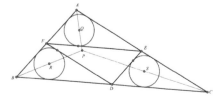

Figure 1.54.

We can take this a step further by considering points *J*, *K*, and *L*, at which lines *AQ*, *BR*, and *CS* intersect the nearest side of the inside triangle *DEF*, as shown in Figure 1.55. Surprisingly, the lines joining the vertices of the inside triangle *DEF* and the points, *J*, *K*, and *L* meet at point *P*. Once again, in one configuration we have found several concurrencies. An ambitious reader may look for further concurrencies.

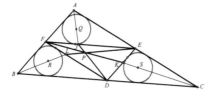

Figure 1.55.

This time we will work in the exterior of a triangle as well as the interior. The centers of the circles, *P*, *Q*, and *O*, are tangent to the extensions of the sides of triangle *ABC*. Such circles are called *escribed circles*, while the circle inside the triangle is called the *inscribed circle*. The points of tangency of the circles with each of the three sides are clearly marked in Figure 1.56. Circle *P* is tangent to the three sides at points *H*, *F*, and *G*. The circle with center *Q* is tangent to the three sides of the triangle at points *R*, *E*, and *N*. And the circle with center *O* is tangent to the three sides of the triangle at points *J*, *D*, and *M*. This configuration gives us lots of concurrent lines:

> *AD*, *BE*, and *CF* are concurrent at point *X*.
> *PC*, *AO*, and *QB* are concurrent at point Y.

We also have a number of collinear points, such as *P*, *A*, and *Q*; *P*, *B*, and *O*; and *Q*, *C*, and *O*.

A motivated reader may want to search for other collinearities or concurrencies in this very rich diagram.

Next, we consider three circles of different sizes, which are not linked to each other except by tangent lines shared by the circles two a time, as shown in Figure 1.57. When we connect the three points of intersection, *R*, *T*, and *S*, of the common tangent lines with the centers of the circles opposite them, we again find a surprising concurrency. Note that the circles were placed randomly. This makes this concurrency all the more amazing!

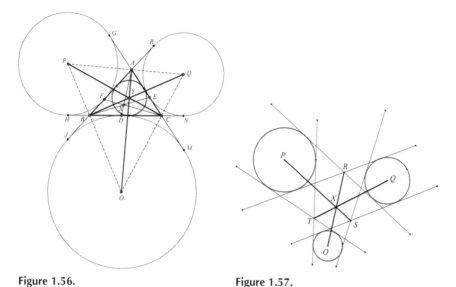

Figure 1.56.   **Figure 1.57.**

SPECIAL TRIANGLES PLACED ON SIDES OF A GENERAL TRIANGLE

A famous relationship involves placement of equilateral triangles on the three sides of a randomly drawn triangle. Napoleon Bonaparte (1769–1821), who was enamored with mathematics, is believed to have discovered this relationship. As shown in Figure 1.58, the lines drawn from each vertex of the original triangle to the remote vertex of the equilateral triangle on the opposite side are concurrent. Note that triangle ABC could be any shape, and this relationship will still hold. Furthermore, these three concurrent line segments are equal in length: $AE = BD = CF$.

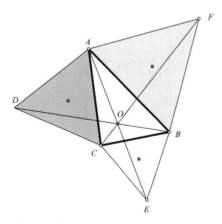

Figure 1.58.

Moreover, the centers of the three equilateral triangles on the sides of triangle *ABC*, when joined by line segments, form another equilateral triangle, as shown in Figure 1.59. This configuration is often called *Napoleon's triangle*.

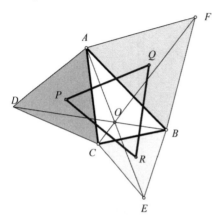

Figure 1.59.

There is still more to be discovered in this Napoleon's triangle. The circumcircles of each of the equilateral triangles are concurrent at point *O*, the point of concurrency of the original three lines. This is shown in Figure 1.60.

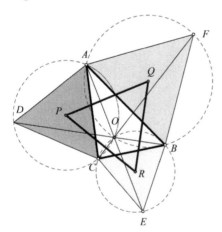

Figure 1.60.

We continue to find more beauty in this noteworthy geometric configuration. The point *O* is called the *equiangular point* of triangle *ABC*, since $\angle AOB = \angle BOC = \angle COA$, as shown in Figure 1.61.

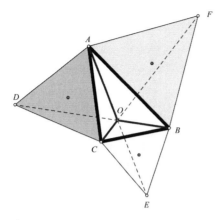

Figure 1.61.

There is one final surprising equilateral triangle in this Napoleon's triangle. All we need to do is to create a parallelogram beginning with segments *AD* and *DC*, which gives us the parallelogram *ADCK*. Lo and behold, we obtain *AKF* as another equilateral triangle, as shown in Figure 1.62.

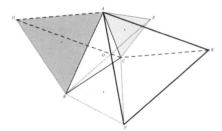

Figure 1.62.

Now that we have had equilateral triangles and circles on each side of a triangle, we construct in a very unusual way a triangle that can be placed on each of the three sides of a given triangle. We will do this by using a process called *reflection*, in which we reflect triangle *ABC* through line *AC* to create triangle *ADC*, as shown in Figure 1.63. The technique for doing this is to draw a line from point *B* perpendicular to *AC* at point *G*, then mark off a length equal to *BG* on the extension of *BG* and call it point *D*. We then have triangle *ADC* as the reflection of triangle *ABC* in the line *AC*.

We now employ this technique two more times. This will yield reflection of triangle *ABC* through the line *AB*, giving us triangle *ABE*. The third time, the triangle *ABC* will be reflected in side *BC*, creating the triangle *FBC*, shown in Figure 1.64.

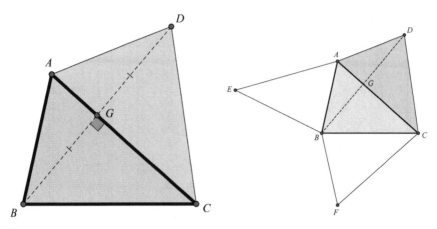

Figure 1.63.   **Figure 1.64.**

Now that we have three triangles, each of which is congruent to the original triangle *ABC*, we will construct their circumcircles, as shown in Figure 1.65. Unexpectedly, the three circles are concurrent. In other words, they share a common intersection point, *P*.

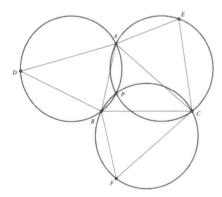

Figure 1.65.

Now if that isn't enough, we can find still another concurrency point, this time, once again, with concurrent lines. In Figure 1.66, the three concurrent lines emerge by joining the center of each circle with the remote vertex of the original triangle *ABC*. We then observe that lines *AS*, *BQ*, and *CR* are concurrent at point *O*.

Suppose we now return to our Napoleon's triangle and use our newly developed skill of reflecting a triangle in a side of another triangle.

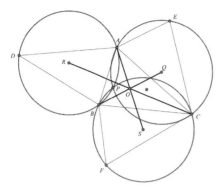

Figure 1.66.

In Figure 1.67, we reflect each of the equilateral triangles in the side on which they are drawn. Note that the reflected triangles (dashed lines) brought with them their center point, which when combined with segments created another equilateral triangle. When we combine the reflected center points of the reflected triangles in Figure 1.67, we once again get an equilateral triangle, $P'R'Q'$.

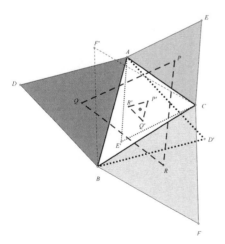

Figure 1.67.

CONCURRENT CIRCLES

Another interesting set of three concurrent circles can be constructed from the triangle partitioned into four congruent triangles by joining the midpoints of the sides of the outside triangle, which we considered earlier in Figure 1.52.

When we draw the circumcircles of the three "outside" triangles, they meet at point *P*, as shown in Figure 1.68.

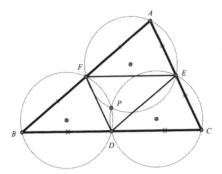

Figure 1.68.

To further add to the wonder of these three concurrent circles, when we draw lines from each of the large triangle's vertices to the centers of the three circles, we find that these lines also are concurrent. Amazingly enough, they are concurrent at the very same point as the three circles. We show this in Figure 1.69.

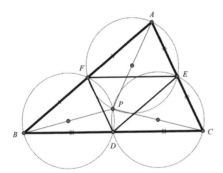

Figure 1.69.

The configuration shown in Figure 1.69 can be generalized by taking any point on each side of triangle *ABC* and constructing three circles, using the vertices as a third point to determine the circles, as shown in Figure 1.70. Notice that the three circles contain a common point, *P*. This point, known as the *Miquel point* of a triangle, is named after the French mathematician Auguste Miquel (1816–1851), who first discovered this wonderful relationship.

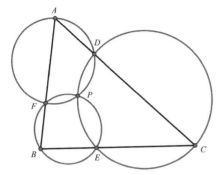

Figure 1.70.

There are some other interesting features about the Miquel point. For example, when line segments are drawn from the Miquel point to the other circle intersection points, which are on the sides of the original triangle, the angles formed with those sides are equal, as shown in Figure 1.71. There you can see that $\angle AEP = \angle CDP = \angle BEP$. A special appreciation of this relationship results when we recall that we began with any triangle, which provides a generalization beyond this one illustration.

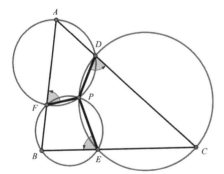

Figure 1.71.

Note that for some triangles, such as obtuse triangles, the point of concurrency, P, of the three circles could be outside the triangle, as shown in Figure 1.72. Yet, the properties we have seen for the acute triangle hold for the obtuse triangle as well.

Further, if we join the center points of the three circles in the Miquel configuration, we amazingly get a triangle that is similar to the original one. That is, in Figure 1.73, triangle ABC is similar to triangle RSQ, since the three corresponding angles are equal, as marked. Of course, this also holds true for the obtuse triangle illustration shown in Figure 1.73.

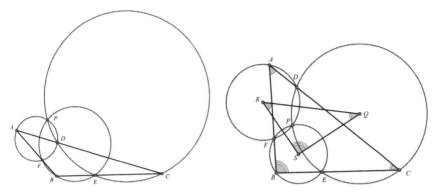

Figure 1.72.   **Figure 1.73.**

We can take this Miquel configuration a step further. Consider any second triangle, where each vertex is on one of the Miquel circles and each side contains one of the three Miquel points. Such a triangle is similar to the original triangle. We have one such configuration in Figure 1.74, where we begin with triangle ABC and then construct triangle GHK so that one vertex is on each of the three circles, and each side contains one of the Miquel points E, D, and F. When we do this, we have constructed, surprisingly, a triangle (GHK) that is similar to the original triangle (ABC).

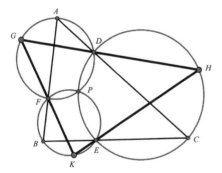

Figure 1.74.

Miquel's theorem is remarkable in that the three points on the sides of the original triangle can also be on the extensions of two sides of the given triangle, as shown in Figure 1.75. In comparison to the configuration shown in Figure 1.70, points F and E are not on the internal segments of the triangle's sides but rather on their extensions. We then follow the Miquel procedure of drawing the three circles, as we have previously done, and notice that they also are concurrent at point P. Of course, all the aforementioned properties will once again hold true. At this point you may wonder if there any limitations to the Miquel theorem. Keep reading!

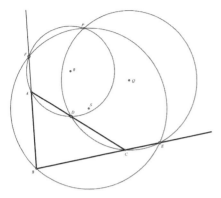

Figure 1.75.

We can even apply Miquel's theorem and concurrent circles to a quadrilateral. If we extend the sides of a quadrilateral until the opposite sides meet—assuming they are not parallel—the resulting configuration is called a *complete quadrilateral*. At the same time, we will have formed four triangles, and for each of these we will apply the Miquel circles. Amazingly, we find that all the circles we draw share one common point of concurrency, *P*. In Figure 1.76, focus on the four triangles △*ABC*, △*ADE*, △*BFE*, and △*CDF* and notice that the four circumcircles meet at point *P*.

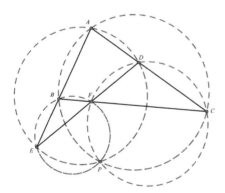

Figure 1.76.

We shouldn't think that three circles cannot create a concurrency without a triangle. As shown in Figure 1.77, we have three randomly drawn circles, each of which is tangent to the larger circle internally, and we mark the points of intersection of the circles as *D*, *E*, *G*, and *F*. When we join the centers of the circles with a remote intersection of the other two circles, we find that they are concurrent at point *P*. That is, *DQ*, *ER*, and *FS* are the three concurrent lines.

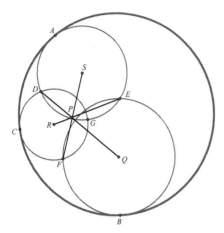

Figure 1.77.

A RECTANGLE CONCURRENCE

Let us now leave circles for a while and concentrate on rectilinear figures. Here, a rather simple construction leads to a most unexpected concurrency. We begin with any rectangle *ABCD*, as shown in Figure 1.78. We then draw a line parallel to the horizontal lines cutting the vertical lines at points *E* and *J*. We do the same thing with a vertical line cutting the horizontal lines at points *F* and *K*. Now comes the most unexpected result: We draw the diagonal of rectangle *ADJE*. Then we draw the diagonal of rectangle *FDCK*. When we draw the diagonal of rectangle *ERKB* and extend it, we find that all three of these diagonals are concurrent at point *P*. Since this can be done for any shape rectangle with any parallel lines, we show in Figures 1.78, 1.79, and 1.80 a few rectangles of different shapes that all yield the same result.

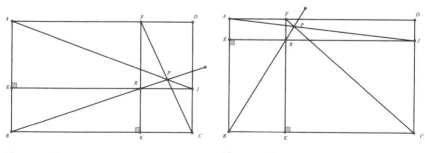

Figure 1.78.   **Figure 1.79.**

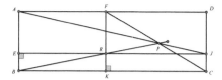

Figure 1.80.

QUADRILATERALS ON THE SIDES OF A TRIANGLE

Another equally surprising concurrency occurs when we take any random triangle (in this case triangle *ABC*, shown in Figure 1.81) and draw squares on two of its sides, as we have done on sides *AB* and *AC*. We then draw a perpendicular from a remote vertex of one of the squares to the furthest side of triangle *ABC*, as we have done with *DK* perpendicular to *BC*. We then do the same thing with the other square to get *FL* perpendicular to *AB* so that these two perpendiculars meet at point *P*. Most unexpectedly, when we draw the altitude from *B* to *AC* of triangle *ABC*, it turns out to be concurrent with the previous two perpendiculars.

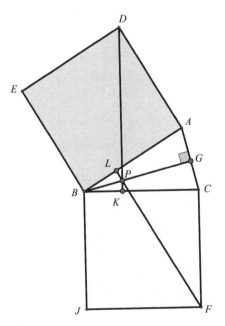

Figure 1.81.

We can extend this initial diagram by adding a congruent square on each of the two existing squares, as shown in Figure 1.82. By joining a remote vertex of the square and triangle twice, we find that they are concurrent with the triangle's altitude from the vertex where the squares meet. In other words, the two lines *TC* and *FA* are concurrent with the altitude *BG* of triangle *ABC*.

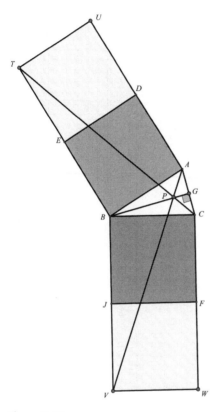

Figure 1.82.

Suppose we now make triangle *ABC* a right triangle with the right angle at vertex *B*. Once again, we will join remote vertices of the triangle and square twice, as shown in Figure 1.83 by line segments *DC* and *AF*. Surprisingly, they intersect the altitude from *B* to the hypotenuse *AC* at point *P*. Again we have three concurrent lines—rather unexpectedly.

This time we construct a square on each side of triangle *DEF* so that the external sides of these three squares, when extended, form triangle *ABC*, as shown in Figure 1.84. When we draw the (extended) lines *AD*, *BE*, and *CF*, we find that they are concurrent at point *P*.

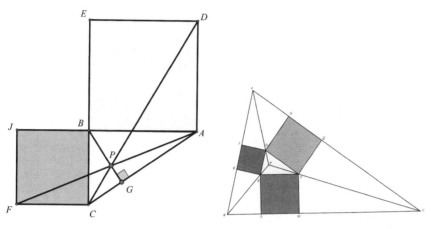

Figure 1.83.   **Figure 1.84.**

While placing squares on the sides of a triangle we can discover a number of other concurrencies. In Figure 1.85, we begin with any randomly drawn acute triangle *ABC*. We can easily locate the center of each of the squares by getting the intersection of its diagonals. When we join the center of a square with the remote vertex of the triangle, we find that those three lines, *AS*, *BQ*, and *CR*, are concurrent. Bear in mind, as with many of the other examples shown here, that these concurrencies are independent of the shape of the original triangle. That is part of the beauty of geometry we are trying to demonstrate.

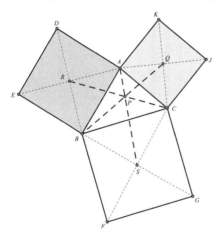

Figure 1.85.

When seeking further concurrencies, we can try some rather inventive ways to find them. Figure 1.86 shows a most unusual arrangement of four lines that have a common point of intersection. Two of the lines, *AF* and *CE*, join triangle vertices with remote vertices on squares on opposite sides of triangle *ABC*. The third line, *DG*, joins the two remote vertices of the squares thus far involved. The fourth line also shares a common point of intersection, *P*, and this line, *BQ*, joins a third vertex of the triangle with the center of the square on the opposite side. This would be a more difficult concurrency to discover on one's own, which might be reason to admire it even more.

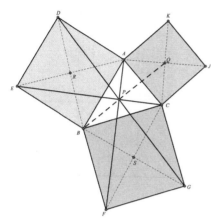

Figure 1.86.

Sometimes we can find concurrencies with part of a configuration, as shown in Figure 1.87, where we ignore square *BCGF* and work with the other two remaining squares. Here we find that lines *EJ*, *BK*, and *DC* are concurrent at point *P*. We could just as easily have ignored one of the other squares and repeated this procedure with the remaining two squares. Therein lies the beauty of the situation!

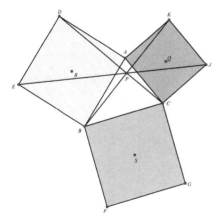

Figure 1.87.

We can discover several more concurrencies in this configuration, where a square is placed on each side of a randomly drawn triangle. Figure 1.88 shows an unusual situation rather than a concurrency. When we draw the line of centers, RQ, and then compare it to the line joining the common vertex of these two squares to the center of the third square, we find that these two lines are not only perpendicular but also the same length; that is, $RQ \perp AS$ and $RQ = AS$. Once again, the beauty lies in the fact that this can be done for any triangle.

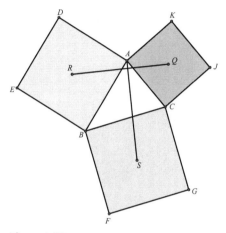

Figure 1.88.

Another concurrency that can be found in this configuration is shown in Figure 1.89. Here, lines EC and JB are concurrent with the perpendicular line from vertex A to side FG.

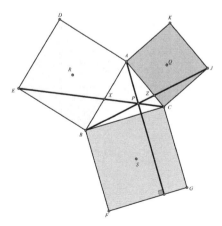

Figure 1.89.

Now, we enhance the diagram shown in Figure 1.89 by constructing parallelograms between each of the squares, so that we have the following parallelograms: *AKXD*, *CJZG*, and *BFWE*. Once again, we locate the centers of each of these parallelograms by drawing the diagonals. When we connect these centers to the centers of the remote squares by drawing lines *NQ*, *MS*, and *RT*, again, to our amazement, the lines are concurrent at point *P*. This is shown in Figure 1.90.

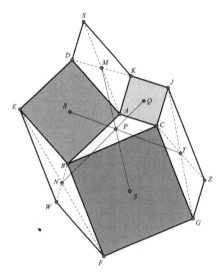

Figure 1.90.

We can establish still another concurrency from the previous diagram (Figure 1.90). This time we connect the center of each square with the remote vertex of the opposite parallelogram. In other words, segments *XS*, *WQ*, and *ZR* are now also concurrent at point *P*, which is shown in Figure 1.91.

Yet another concurrency can be established in this configuration (Figure 1.92). We can join midpoints *U*, *V*, and *Y* of the remote sides of the squares, *ED*, *KJ*, and *FG* respectively, with the opposite remote vertex of each of the parallelograms so that lines *UZ*, *XY*, and *VW* are concurrent.

There are even more concurrencies that we can identify in this configuration. This time we connect the midpoints of the sides of the triangle to the center point of the opposite parallelograms. In Figure 1.93, we show that lines *ML*, *HT*, and *NI* are concurrent at point *P*.

While we are still on this configuration, let's draw an altitude from each vertex of the original triangle to each of the parallelograms' diagonals. This results in another concurrency. When *NB*, *TC*, and *MA* are extended, they meet at point *P*, as shown in Figure 1.94.

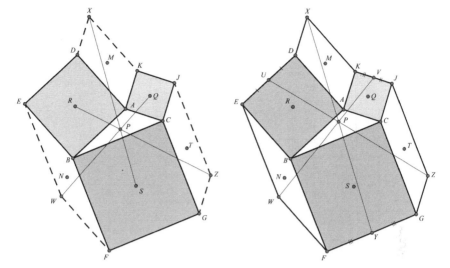

Figure 1.91.   **Figure 1.92.**

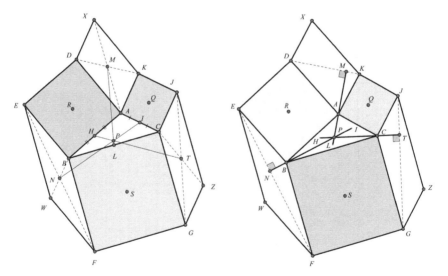

Figure 1.93.   **Figure 1.94.**

We find yet more concurrent lines by taking the perpendicular bisector of each of the diagonals of the parallelograms, as shown in Figure 1.95, where M, N, and T are midpoints. We then have the following concurrency: HM, IT, and LN.

Besides the various concurrencies that exist in this configuration, there are also equalities to be found. We offer one here and leave the others for the

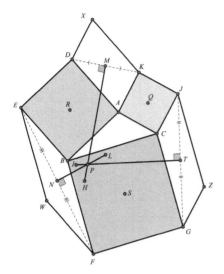

Figure 1.95.

reader to discover. In Figure 1.96, we notice that $AW = AZ$. Good luck in your search for the other equalities!

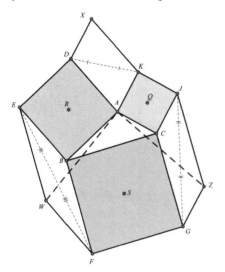

Figure 1.96.

MORE PLACEMENTS OF SQUARES

We now explore placement of squares onto a randomly drawn quadrilateral, $TLUV$, shown in Figure 1.97. First, we will join the center points of opposite squares so that we have lines *YZ* and *XW* meeting at point *P*. Curiously enough, when we join the midpoints of the four lines joining the squares—*J* the midpoint of *AH*, *K* the midpoint of *GF*, *N* the midpoint of *DE*, and *M* the midpoint of *BC*—we find that lines *MK* and *NJ* also meet at point *P*. In effect we have four lines, which are all concurrent at point *P*.

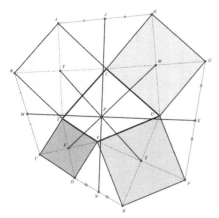

Figure 1.97.

Having placed squares on triangles and then on a quadrilateral, we now place squares on a point, as shown in Figure 1.98. The three squares are shown with the sole consideration that they all share the common point *P*. We then join vertices of adjacent squares with the three line segments *AK*, *GF*, and *CD*. As shown in Figure 1.98, we join the midpoints of those three lines with the opposite square's center. Amazingly, these three lines, *YM*, *XQ*, and *NZ*, are always concurrent (at point *R*) regardless of the size of the squares and their placement as long as they share a common vertex with the others. This is surely an example to be cherished!

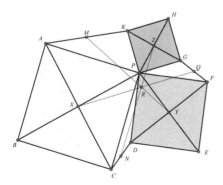

Figure 1.98.

We can even find concurrencies where two squares share a common vertex point, *P*. In Figure 1.99, we notice two randomly placed squares of different sizes that share the common vertex *A*. When we draw the lines *BE*, *CF*, and *DG*, we notice that they are concurrent at point *P*. As with many of our other examples, placement of the two squares will not affect the concurrency.

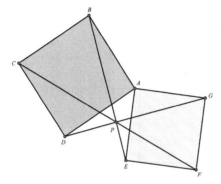

Figure 1.99.

To show that the placement of the two squares will not affect the concurrency, we offer Figure 1.100. Here the squares have changed size and position, and still, the concurrency remains intact.

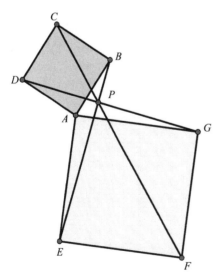

Figure 1.100.

There are also configurations where analogous polygons are embedded on the sides of a triangle. Consider, for example, Figure 1.101, where we begin with triangle *ABC* and select any point *P*. We then join *P* to the three vertices and create parallelograms, as shown in the figure. When we connect each vertex of the original triangle to the remote vertex of the parallelogram embracing the opposite side, we find that lines *AF*, *BE*, and *CD* are concurrent at point *R*. For the ambitious reader we offer a small enhancement to this already amazing concurrency: $AF^2 + BE^2 + CD^2 = (AB^2 + BC^2 + AC^2) + (AP^2 + BP^2 + CP^2)$.

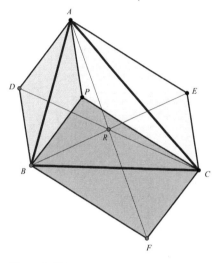

Figure 1.101.

We can also find surprising concurrencies in two regular pentagons placed *in any way we choose*, except that they share one common vertex. In Figure 1.102, that common vertex is point *X*. When we join the corresponding vertices of the two pentagons with lines *AE*, *BF*, *CG*, and *DH*, as shown in Figure 1.102, we find that these four lines are concurrent at point *P*, which could be anywhere, depending on the relative sizes and placement of the pentagons.

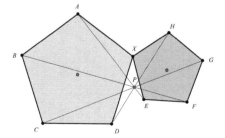

Figure 1.102.

We can extend this scheme further by repeating it for hexagons in place of the pentagons we previously used and still find a concurrency. In Figure 1.103 we have two hexagons that share a common vertex N, yet are located randomly and are of different sizes. When we join the points consecutively, we find that the five lines AF, BG, CH, DJ, and EK are concurrent at point P.

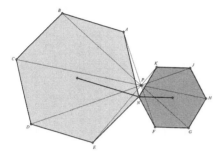

Figure 1.103.

BACK TO TRIANGLE PLACEMENTS

In our pursuit of further concurrencies, let us consider the configuration shown in Figure 1.104. Here we place square $DEFG$ inside triangle ABC in such a way that two sides of the square are parallel to altitude AH. Once again, we find an unexpected concurrency when we draw lines BFJ and CGK, which are concurrent with altitude AH at point P. Again, the beauty lies in the fact that it is the placement, not the size, of the triangle and the square that is important.

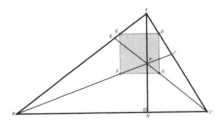

Figure 1.104.

Here is a rather unusual arrangement for those accustomed to working with related triangles. Consider two noncongruent triangles placed one inside the other, with the corresponding sides parallel, as shown in Figure 1.105. Here, the sides of triangle ABC are parallel to the corresponding sides of triangle DEF. We can clearly see that the lines joining the corresponding vertices are concurrent at point P.

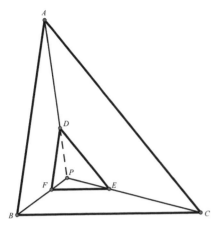

Figure 1.105.

Another way of placing triangles within other triangles is where two triangles are placed in such a way that lines from each vertex of the larger triangle are perpendicular to the nearer side of the inside triangle. Again, it turns out that these lines are concurrent.

This configuration is shown in Figure 1.106, where each of the vertices of triangle DEF are on the sides of triangle ABC, and the lines from vertices A, B, and C are each perpendicular to sides DF, DE, and EF at points H, G, and J, respectively. These three perpendicular lines, AH, BG, and CJ, are concurrent at point R.

If this isn't impressive enough, we can take it one step further and show another concurrency for this configuration. When we erect perpendiculars at points D, E, and F to each of the sides of triangle ABC, they will be concurrent at point P (Figure 1.106).

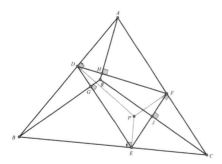

Figure 1.106.

Our next example shows how two triangles inscribed in the same circle have a relationship through their angle bisectors and altitudes. Figure 1.107 shows triangle ABC and its circumcircle O. We then draw the angle bisectors of each of the angles of triangle *ABC* to meet the circumcircle at points *D*, *E*, and *F*. It turns out that the point of intersection, *I*, of the angle bisectors also serves as the orthocenter of triangle *DEF*. In other words, point *I* is also the point of intersection of altitudes *DX*, *EY*, and *FZ* of triangle *DEF*. Therefore, we can say that the two triangles are related by sharing lines that are respectively angle bisectors and altitudes.

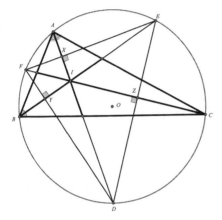

Figure 1.107.

Although the next configuration is complex, it again reveals concurrency in a way you might least expect. We begin with triangle *ABC* and its inscribed circle *O*, as shown in Figure 1.108. We then randomly select any diameter of circle *O*, and from each vertex of triangle *ABC* we draw a perpendicular intersecting the diameter at points *D*, *E*, and *F*. From these three points, *D*, *E*, and *F*, we then draw another set of perpendiculars to each of the three sides of the triangle, *BC*, *AC*, and *AB*, respectively, intersecting the sides of the triangle at points *P*, *Q*, and *R*. Unexpectedly, we find that *EQ*, *FR*, and *DP* are concurrent at point *X*. This configuration, once again, points out the beauty of geometry, which sometimes can be achieved through less attractive beginnings.

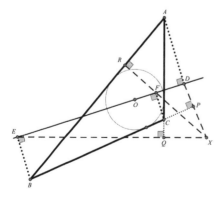

Figure 1.108.

Our next discovery of concurrency likewise is complicated. We begin with triangle *ABC*, which is shown in Figure 1.109, and select any point somewhere inside the triangle. We then draw a line *l* containing point *P* and have it intersect the sides of the triangle at points *X*, *Y*, and *Z*. We now let extended lines *AP*, *BP*, and *CP* intersect the circumcircle of triangle *ABC* at points *R*, *S*, and *T* respectively. Unexpectedly, lines *RX*, *SZ*, and *TY* are concurrent at point *Q*. This elaborate configuration shows how one concurrency creates another that appears to be completely unrelated and where we would least expect it.

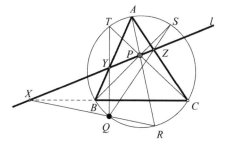

Figure 1.109.

POLYGON CONCURRENCIES

Let us now focus on some polygon concurrencies. In Figure 1.110 we see a hexagon circumscribed about the circle. If this were a regular hexagon, the diagonals would certainly be concurrent. But here we have a randomly drawn nonregular hexagon with the sole condition that it is circumscribed about a circle—meaning that the circle is tangent to each of the six sides of the hexagon. Unexpectedly, in this case, we once again find that the diagonals are concurrent at point *P*.

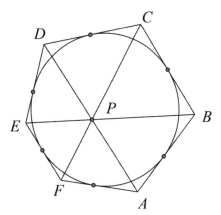

Figure 1.110.

This unusual relationship was discovered in 1806 by a 21-year-old French student, Charles Julian Brianchon (1783–1864), who later became a professor of mathematics. Furthermore, it holds true not only for a circle but also for an ellipse. That is, if we had a hexagon circumscribed around an ellipse, the same thing would be true: the diagonals joining opposite vertices would be concurrent, as shown in Figure 1.111.

Figure 1.111.

While we're still on the topic of polygons, let's consider a nonregular pentagon circumscribed about a circle, as shown in Figure 1.112. Here, two diagonals, *AD* and *BE*, intersect at point *P*. Then the line joining vertex *C* to the point of tangency *F* on the opposite side of the pentagon will be concurrent with the other two lines at point *P*. There is a very subtle relationship between this situation and that of the hexagon. We leave its discovery to the ambitious reader.

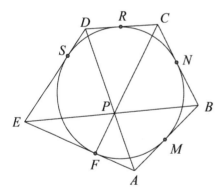

Figure 1.112.

A lot more concurrencies can be found on polygons. For entertainment and as a challenge to the motivated reader we begin by considering an 18-sided regular polygon (i.e., a polygon that has equal sides and equal angles), as shown in Figure 1.113. A host of surprising concurrencies appear in the next few figures, which will further allow us to appreciate them. We describe some of these examples here and leave the others for the reader to discover. Beginning with Figure 1.113, five lines can be symmetric about the diagonal *AK*, which are all concurrent at point *P*.

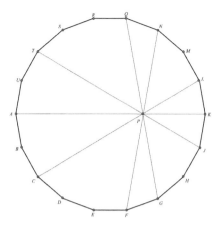

Figure 1.113.

In Figure 1.114 we consider an 18-sided regular polygon with four diagonals symmetric around the center diagonal *AK*, once again concurrent at point *P*.

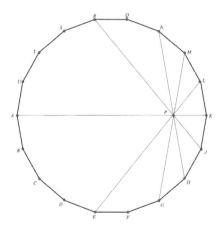

Figure 1.114.

Figures 1.115 to Figure 1.118 show several other concurrencies that can be found in the 18-sided regular polygon.

Before we challenge the reader to find these other concurrencies, we describe one more. In it we have five lines, shown in Figure 1.118, that are concurrent at point *P*. Notice that a certain symmetry exists among the line segments in the examples that we provided.

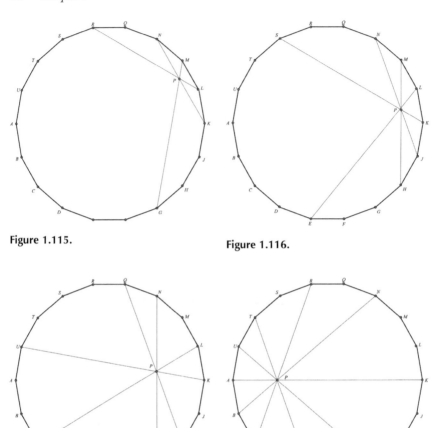

Figure 1.115.

Figure 1.116.

Figure 1.117.

Figure 1.118.

We are now finished with the concept of concurrency of lines. We move on to the analogue: the collinearity of three or more points, which are points on a straight line.

2

Collinearity

Concurrency of lines and collinearity of points are analogs and in some cases closely related. One of the better known relationships between these two concepts was demonstrated by French mathematician Gerard Desargues (1591–1661). This relationship relates concurrency to collinearity, and vice versa.

DESARGUES' THEOREM AND BEYOND

We begin by placing two triangles in such a way that the lines connecting the corresponding vertices will be concurrent. According to Desargues' theorem, once this is achieved the pairs of corresponding sides will then meet in three collinear points. In Figure 2.1 the corresponding vertices are A_1 and A_2, B_1 and B_2, and C_1 and C_2. When we connect these vertices with lines we notice that A_1A_2, B_1B_2, and C_1C_2 meet at point P. When we extend the corresponding sides, C_1B_1 and C_2B_2 meet at point A', A_1B_1 and A_2B_2 meet at point C', and A_1C_1 and A_2C_2 meet at point B'. We find that these three points, A', B', and C', are collinear.

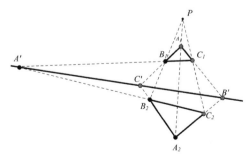

Figure 2.1.

Here we can see how collinearity and concurrency relate. To further strengthen this point, we could have pursued this configuration conversely by placing the two original triangles in such a way that the corresponding sides connected through their extensions meet in three collinear points. Then the lines joining the corresponding vertices would be concurrent.

This amazing relationship that Desargues offered the world of mathematics allows us to appreciate other unexpected relationships—in this case involving concurrency and collinearity. We will consider one of these in Figure 2.2, where the points of tangency of the inscribed circle in triangle *ABC* are *M*, *N*, and *L*. When we connect the corresponding vertices of triangles *ABC* and *LMN*, recalling the Gergonne point (see page 17) enables us to realize that line segments *AL*, *BM*, and *CN* are concurrent. This relationship can be easily established by Ceva's theorem, noting that two tangent segments from an external point to the same circle are equal (that is, *AM* = *AN*, *BN* = *BL*, and *CM* = *CL*). Thanks to Desargues' theorem, we see that the line extensions of the corresponding sides of the two triangles, *ABC* and *MNL*, meet at three collinear points, *P*, *Q*, and *R*, as shown in Figure 2.2, where *MN* and *CB* meet at point *P*, *LM* and *BA* meet at point *R*, and *LN* and *CB* meet at point *Q*.

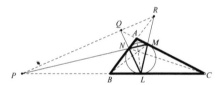

Figure 2.2.

By this line of reasoning, we can apply an apparently analogous situation, shown in Figure 2.3, since we have already established that the altitudes of a triangle are concurrent. As in the previous example, the lines joining the vertices of triangles *ABC* and *LMN* are concurrent. Once again invoking Desargues' theorem, we find that the line extensions of the corresponding sides meet at three collinear points (*P*, *Q*, and *R*): *MN* and *CB* at point *P*, *LM* and *BA* at point *R*, and *LN* and *CB* at point *Q*.

Now that we have some experience with Desargues' theorem, we embark toward a truly unexpected result that we can justify very nicely by applying it. As shown in Figure 2.4, we select points *E*, *F*, *G*, and *H* on the sides of parallelogram *ABCD* so that lines *GH*, *AC*, and *EF* are concurrent at point *P*. Unexpectedly, when we draw lines *HE*, *DB*, and *GF*, they too are concurrent at point *Q*.

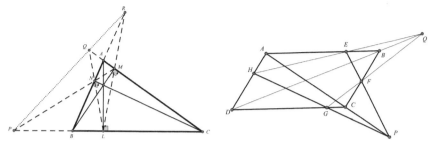

Figure 2.3.   **Figure 2.4.**

This surprising result is even better appreciated when we see how simply it is justified by Desargues' theorem. To do that, consider triangles *DHG* and *BEF* (highlighted in Figure 2.5). Our original setup had these two triangles placed so that their corresponding sides met at the collinear points *A*, *C*, and *P*. By Desargues' theorem this tells us that the lines joining the corresponding vertices *HE*, *DB*, and *GF* are concurrent at point *Q*. We might consider this a surprising application of Desargues' theorem!

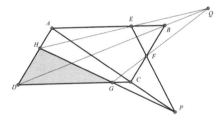

Figure 2.5.

UNEXPECTED SURPRISES FROM SIMSON'S THEOREM

As shown in the previous example, concurrency of lines is analogous to collinearity of points (more than two points that lie on the same straight line). When considering collinear points involving triangles, Simson's theorem comes into play.

We should appropriately credit the originator of this theorem, since it touches upon one of the great injustices in the history of mathematics. This theorem was originally published by English mathematician William Wallace (1768–1843) in Thomas Leybourn's *Mathematical Repository* (1799–1800). Through careless misquotes, the theorem has been attributed to Scottish mathematician Robert Simson (1687–1768), whose edition of Euclid's *Elements* was long the basis for the study of geometry in the English-speaking world and, more specifically, greatly influenced American high school geometry courses.

Simson's theorem states that the feet of the perpendiculars drawn from *any* point on the circumscribed circle of a triangle to the sides of the triangle are collinear. This is shown in Figure 2.6, where *P* is any point on the circumscribed circle of triangle *ABC*. We then draw *PY* perpendicular to *AC* at *Y*, *PZ* perpendicular to *AB* at *Z*, and *PX* perpendicular to *BC* at *X*. According to Simson's theorem, points *X*, *Y*, and *Z* are collinear. The line that contains these points is usually referred to as the *Simson line*.

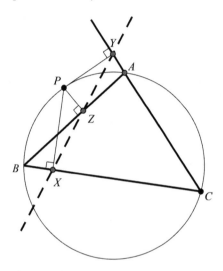

Figure 2.6.

A curious phenomenon occurs when we construct the Simson line from the intersection point on the circumscribed circle and the extension of one of the triangle's altitudes. This Simson line is parallel to the tangent line at the vertex from which this altitude emanates. Figure 2.7 shows an example. When altitude *BD* (emanating from point *B*) of triangle *ABC* meets the circumscribed circle at point *P*, then the Simson line of triangle *ABC* with respect to *P* is parallel to the line tangent to the circle at *B*.

Another interesting property of the Simson line is that it bisects the line that joins the orthocenter with the generator point of the Simson line. We can see this in Figure 2.8, where point *P* is used to construct the Simson line, *XZY*, of triangle *ABC*. Line *PH*, joining the orthocenter, *H*, of the triangle with point *P*, is bisected by the Simson line at point *M*, or *PM* = *HM*.

Another curiosity here is that if two Simson lines are constructed for the same triangle by two distinct points on the circumscribed circle, the angle formed by the Simson lines is half the measure of the arc the two points intercept on the circle. In Figure 2.9, Simson lines *YZX* and *UVW* intersect to form angle *MTN*, which is half the measure of arc *PQ*.

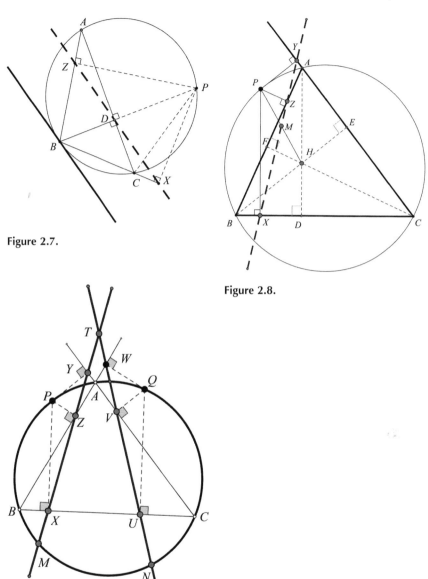

Figure 2.7.

Figure 2.8.

Figure 2.9.

Often, as we have seen so far, three circles can share a common point of intersection. With the support of Simson's theorem, however, we can justify the amazing relationship that four circles share a common intersection point, as shown in Figure 2.10. Here four lines, *AB*, *BC*, *EC*, and *ED*, have created four triangles,

ABC, *FBD*, *EFA*, and *EDC*. The circumcircles for these triangles all contain a common point, *P*. The dashed lines in the figure show the ambitious reader how Simson's theorem helps us guarantee the concurrency of the four circles.

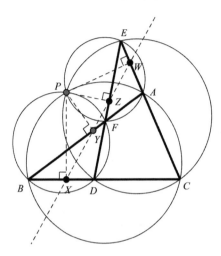

Figure 2.10.

Consider three Simson lines drawn for the same triangle, shown in Figure 2.11. The Simson lines of triangle *ABC*—using points *P*, *Q*, and *R*— form triangle *NST*. When we compare triangle *NST* to *PQR*, the triangle created by the three points on the circumcircle generating the Simson lines, we find that triangle *PQR* is similar to triangle *NST*. This is quite an amazing leap for Simson lines.

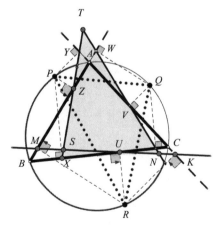

Figure 2.11.

Another concurrency develops after we draw the three altitudes, *AD*, *BE*, and *CF*, of triangle *ABC*, as shown in Figure 2.12. Connecting the feet of the altitudes generates three collinear points, *P*, *Q*, and *R*. We see this when we draw *FE* to intersect *BC* at point *P*, *CA* to intersect *FD* at point *Q*, and *BA* to intersect *DE* at point *R*.

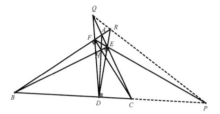

Figure 2.12.

EVEN POLYGONS CAN GENERATE COLLINEARITY

The next few examples consider polygon configurations that lead to collinearity of points. The collinearity here is sometimes well camouflaged. But that is all part of the beauty of geometry, which will be further supported through polygonal examples. We begin with a hexagon, as shown in Figure 2.13.

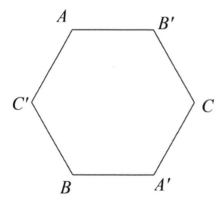

Figure 2.13.

Suppose we consider the vertices, *AB'*, *CA'*, and *BC'* (Figure 2.13), located alternately on two lines (see Figure 2.14). Then suppose we draw the lines that were previously the opposite sides of the hexagon:

AB' and *A'B*; and note their point of intersection *C''*
BC' and *B'C*; and note their point of intersection *A''*
AC' and *A'C*; and note their point of intersection *B''*

We find that the three points of intersection, A'', B'', and C'', of these pairs of opposite sides are collinear. This surprising result was first published in about 300 CE by Pappus of Alexandria in his *Mathematical Collection*.

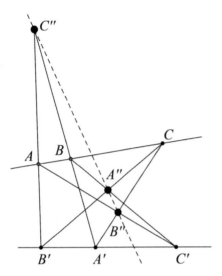

Figure 2.14.

A curious relationship occurs in a hexagon that has no opposite sides parallel and is inscribed in a circle. When the opposite sides are extended until they meet, the three points of intersection will all lie on the same line (are collinear).

This is shown in Figure 2.15, where opposite sides AF and DC meet at point N, EF and BC meet at point M, and ED and AB meet at point L. Like the previous hexagonal relationship, this one was discovered by a soon-to-be famous French mathematician, Blaise Pascal (1623–1662), who published it at age 16. Also similar to the previous case, this relationship holds true not only for circles but also for ellipses, as shown in Figure 2.16.

We now perform a rather unusual procedure. Recall from Figure 2.13 the opposite sides of the hexagon: AF is opposite CD, AF is opposite CD, and AF is opposite CD. We now place these points randomly on a circle, as shown in Figure 2.17, and we notice the following:

AF and CD intersect at point N
BC and FE intersect at point M
AB and ED intersect at point L

Looking at the opposite sides as we did before, we see that, assuming they are not parallel, they are able to intersect. Once again, to our surprise (and awe),

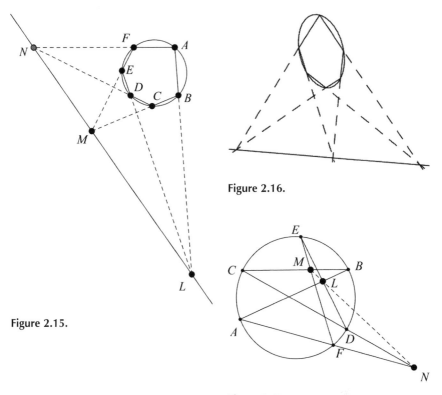

Figure 2.16.

Figure 2.15.

Figure 2.17.

the result is three collinear points. We show this in the carefully constructed version of Figure 2.17, where points *N*, *L*, and *M* are, in fact, collinear.

Collinearity sometimes occurs in surprising and somewhat artificial ways. One such example is shown in Figure 2.18. Here, we see triangle *ABC* and the midpoints of its sides marked with points *M*, *N*, and *L*. We now choose a random point, *P*, somewhere inside triangle *ABC*. From the vertices of the triangle we draw lines through point *P* that intersect the opposite sides at points *D*, *E*, and *F*. Up to this point we have not done anything terribly unusual. Now, however, we do something that will lead to our collinearity, and it may look somewhat artificial—but it's correct! We will join each of the midpoints, *M*, *N*, and *L*, with each of the previously determined endpoints, *F*, *D*, and *E*, respectively. That produces lines *FM*, *ND*, and *EL*, which meet the (extended) respective third sides, *AC*, *AB*, and *CB*, at points *X*, *Y*, and *Z*, respectively. And, as we notice in Figure 2.18, these three points are collinear. Although the construction of this configuration was simple, albeit unusual, it led to an unexpected collinearity.

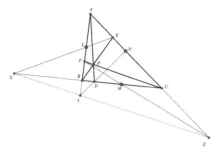

Figure 2.18.

MORE UNEXPECTED COLLINEARITIES

We now embark on a somewhat simpler illustration of collinearity in geometry. Once again, we begin with triangle *ABC* and its circumcircle *O*. We then draw the tangents to the circumcircle at each of the three vertices of the triangle. It turns out that these tangents meet the opposite sides at three collinear points *J*, *K*, and *L*. But as shown in Figure 2.19, the sides of the triangle in each case needed to be extended to meet the tangent lines.

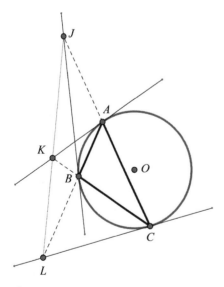

Figure 2.19.

Circles also can produce collinear points, as shown in the following example. We begin with circle *O* (Figure 2.20) and simply draw chords *AB*, *AC*, and *AD* emanating from one randomly selected point, *A*, on the circle.

Next, we construct new circles using each of these chords as the diameters. We then identify the three points of intersection of each pair of circles; circles *P* and *Q* meet at point *X*, circles *P* and *R* meet at point *Y*, and circles *Q* and *R* meet at point *Z*. A simple visual inspection shows that these three points of intersection, *X*, *Y*, and *Z*, are collinear. So we see that collinearity is not limited to straight-line figures. To convince yourself that this is true, you might choose to draw it using dynamic geometry.

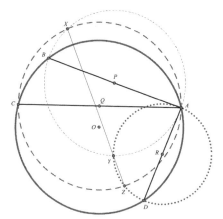

Figure 2.20.

It is particularly interesting that the bisectors of the exterior angles of a nonisosceles triangle meet the opposite sides of the triangle in three collinear points. Consider triangle *ABC* shown in Figure 2.21, where the bisectors of the three exterior angles are *AN*, *BL*, and *CM*. It is then clear that the three points of intersection, *L*, *M*, and *N*, of these exterior-angle bisectors with the extended sides of the triangle are collinear.

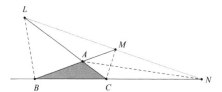

Figure 2.21.

We can take this another step further by considering the angle bisectors of both the interior and exterior angle bisectors of two angles of a triangle. Once we draw these four angle bisectors, we go to the third vertex of the triangle and from there draw the perpendiculars to each of them. When we do this, lo

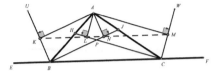

Figure 2.22.

and behold, we once again have collinearity of four points. This is shown in Figure 2.22, where we have the interior angle bisectors *BJ* and *CH* of angles *ABC* and *ACB*, respectively. These angles of triangle *ABC* have exterior angle bisectors *BU* and *CW*. From the third vertex, *A*, of triangle *ABC*, we draw the four perpendiculars meeting each of the angle bisectors at points *K*, *L*, *N*, and *M*, which, as you can see, are collinear. As with most of what we show, this is true for all triangles, where we have four distinctive points.

3

Circles and Concyclic Points

To this point in our journey through the unusual and unexpected relationships that we can admire in geometry we have concentrated on concurrent lines as well as concurrent circles; that is, circles that share a common point, analogous to concurrent lines. We now focus on points that lie on the same circle. Any three noncollinear points will always lie on the same unique circle. Therefore, when we speak of concyclic points, we speak of more than three points that lie on the same circle.

Determining a circle in a rather unusual way once again demonstrates the amazing relationships in geometry. One such example begins with a rectangle, where the width is one-third as long as the length. In Figure 3.1 we show a rectangle where $AD = AM = MN = NB$. We then draw line segment MC meeting diagonal DB at point P. This simple arrangement leaves us with four points, C, B, N, and P, all lying on the same circle. Having four concyclic points is already noteworthy. And it is a simple way for us to begin our journey through discovering more than three points lying on the same circle.

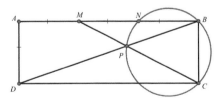

Figure 3.1.

Sometimes concyclic points turn up in quite unusual circumstances. Consider the nest of squares shown in Figure 3.2. This configuration begins with square $ABCD$, then EG and FH are drawn perpendicular to each other at point O, which is the intersection of the diagonals of square $ABCD$. It then turns out that $EFGH$ also is a square, and when the intersection points of the sides of this square are joined with the diagonals of square $ABCD$, we get another square, $MLKN$. The fact that $FO = EO$, and $\angle FAC = \angle EAC = 45°$, allows us to conclude that points A, E, O, and F are concyclic by using the converse—that equal inscribed angles intercept equal arcs on a circle, which in turn generate equal chords.

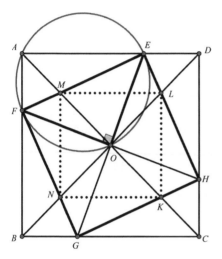

Figure 3.2.

When additional points lie on the same circle, it can be even more note-worthy. Swiss mathematician Leonhard Euler in 1765 first showed that there are six points that lie on the same circle. He found that the midpoints of the sides and the feet of the altitudes of a triangle must lie on the same circle. This is shown in Figure 3.3, where the midpoints of the sides of triangle ABC are points D, E, and F, and the feet of the altitudes are X, Y, and Z, all lying on the same circle with center O.

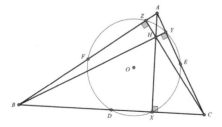

Figure 3.3.

In 1820, two French mathematicians and technical experts in Napoleon's army, Charles-Julian Brianchon (1783–1864) and Jean-Victor Poncelet (1788–1867), discovered three additional points that lie on the very same circle. These three points are the midpoints of the segments joining the orthocenter (the point of intersection of the altitudes) and the feet of the altitudes. Figure 3.4 shows these as points K, L, and M so that $HK = BK$, $HL = LH$, and $HM = CM$. This constitutes the famous *nine-point circle*, also often referred to as the *Feuerbach circle*. This configuration is named after Karl Wilhelm Feuerbach (1800–1834) who in 1822 published a paper that included this relationship and others as well.

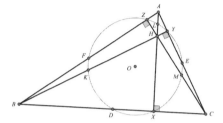

Figure 3.4.

There is still more to admire in this configuration. In Figure 3.5 we draw the line, HP, joining the orthocenter and the center of the circumscribed circle of triangle ABC. It turns out the center O of the nine-point circle is the midpoint of HP, where P is the center of the circumscribed circle of triangle ABC.

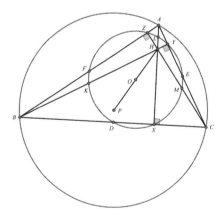

Figure 3.5.

When we draw the three medians of triangle ABC, we locate the centroid, G (the center of gravity of the triangle), which just happens to lie on line HP. But it also is at a trisection point of HP, so that $HG = 2PG$, as shown in

Figure 3.6. This unique line, which exists in all triangles except an equilateral triangle, where all points would mesh in one, is called the *Euler line* and contains four important points: the center of the circumscribed circle, the orthocenter, the centroid, and the center of the nine-point circle.

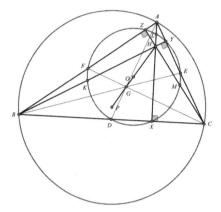

Figure 3.6.

Many other relationships can be found in this configuration; for example, the radius of the nine-point circle of triangle ABC is one-half that of the circumcircle of triangle ABC. In Figure 3.7 we can see that $MO = \dfrac{1}{2}BD$. Furthermore, an ambitious reader might want to verify through geometric constructions that all triangles inscribed in a given circle and having a common orthocenter also have the same nine-point circle. There are also other points of interest on the Euler line, such as the Exeter point, discovered by students at Phillips Exeter Academy in 1986. The Exeter point can be found by extending

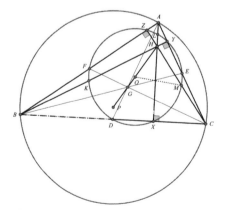

Figure 3.7.

the medians of the triangle to meet the circumcircle of triangle *ABC* at points Q, R, and S. Triangle *TRL* is formed by the tangents to the circumcircle of triangle ABC. Appropriately joining the points Q, R, and S to the vertices of triangle TRL, the Exeter point will be determined. (We leave the diagram to the reader.)

Geometric figures sometimes can be challenging to fully comprehend. Take, for example, Figure 3.8, where we have shaded three triangles, namely, triangles *AHC*, *AHB*, and *BHC*. These three triangles along with the large triangle *ABC* create what is called an *orthocentric system*. In this system, the four points used—*A*, *B*, *C*, and *H*—are each the orthocenter of the triangle formed by the other three points. The surprising result is that these four triangles all share the same nine-point circle. This may not be too easy to visualize at first. Since we already know the nine-point circle for the original triangle *ABC*, let's identify the nine points for one of the other triangles, for example, *AHC*. First we have the midpoints, *K*, *M*, and *D*, of its sides. Now we can locate the feet, *X*, *Y*, and *Z*, of the three altitudes. The midpoints of the segments (*AH*, *BH*, and *CH*) joining the vertices with the orthocenter (*A*) are points *K*, *L*, and *M*. All of these nine points lie on the same nine-point circle. Triangle *XYZ*, which is formed by the feet of the altitudes of triangle *ABC*, is called the *pedal triangle* of the original triangle *ABC*. We will be using the pedal triangle going forward.

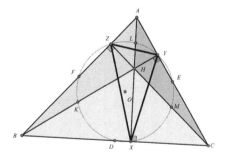

Figure 3.8.

We know that any three noncollinear points determine a unique circle. How surprising it is to find that, surrounding our nine-point circle, are two congruent circles determined as follows: one circle is the circumcircle of the original triangle—triangle *ABC* in Figure 3.9. The second circle is the one containing two vertices of triangle *ABC* and the orthocenter, *H*. That means four congruent circles could be drawn from this configuration. For clarity, however, we show only the circle containing vertices *B* and *C* and the orthocenter, *H*.

In Figure 3.10 we show the four equal circles mentioned above, where we did not want to complicate the original diagram. We can use the pedal triangle

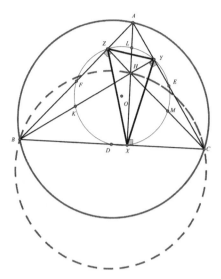

Figure 3.9.

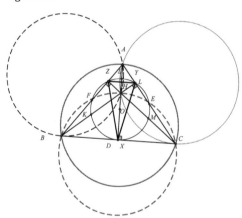

Figure 3.10.

to create a concurrency. In Figure 3.11, XYZ is the pedal triangle of ABC, since it is formed by joining the feet of the altitudes of triangle ABC. Next, we draw lines from each vertex perpendicular to the nearest side of the pedal triangle. Notice that these three lines are concurrent at point P.

Another curiosity in this configuration is that the area of triangle ABC is equal to the product of the radius of the circumscribed circle and half the perimeter of the pedal triangle. It turns out that P is also the center of the circumscribed circle. Therefore, BP is a radius and so:

$$Area \triangle ABC = (BP)\left(\frac{1}{2}\right)(XY + YZ + ZX).$$

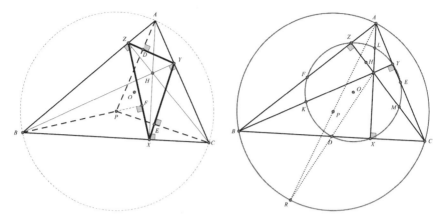

Figure 3.11.   **Figure 3.12.**

Other interesting relationships can be found in this rich configuration generated originally through the discovery of the nine-point circle. For example, for any triangle, the line joining the orthocenter with the midpoint of one of the sides meets the endpoint of the diameter of the circumcircle, which emanates from the vertex, which is opposite the side from which the previously mentioned midpoint was selected. In Figure 3.12, we can see this unexpected concurrence by drawing the line segment joining orthocenter *H* and midpoint *D* of side *BC*, and then drawing the diameter, *APR*, of the circumcircle containing vertex *A*. These two lines meet on the circumcircle at *pedal* point, *R*. This can certainly be applied to each of the three sides of the triangle. It is curious that this intersection point is always on the circumcircle!

Still more can be found in the nine-point circle. The following seemingly contrived example should demonstrate that there are always more novelties to be found within a geometric configuration. Figure 3.13 shows once again the nine-point circle of triangle *ABC*, with the orthocenter designated by *O*. After constructing the bisector of angle *BAC*, which intersects side *BC* at point *T*, we erect a perpendicular from *O* to *AT* and designate it as point *R*. What we now notice is an unexpected collinearity, where points *R*, *P*, and *D* all lie in the same line.

We close our discussion with a rather amazing relationship that the nine-point circle shares with the other circles of a triangle. Consider three circles that are tangent to the three sides of the triangle and lie outside the triangle; that is, circles that are tangent to two extended sides and a third side of the triangle. These are called *excircles* (or escribed circles) of a triangle. The fourth circle that we will consider here is the inscribed circle of the triangle. The relationship here is that the nine-point circle of the triangle is tangent to each

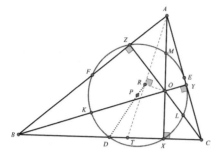

Figure 3.13.

of these four circles. This is shown in Figure 3.14, where circles P, Q, and R are the escribed circles of triangle ABC and circle I is the inscribed circle. Notice that the bold-line circle in Figure 3.14, which is the nine-point circle, is tangent to each of the other four circles. This is also frequently referred to as the *Feuerbach theorem*. It is, of course, true for all triangles as usual!

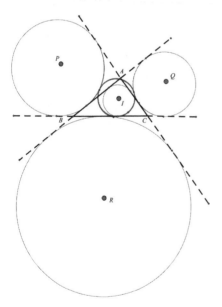

Figure 3.14.

4

On Quadrilaterals

One surprising relationship referring to quadrilaterals is very simply stated and easily proved. In the spirit of this book, however, we merely present it for its beauty and show how it helps us better understand geometric relationships. (We have already encountered it in the introduction of the book.) We begin with any "ugly" quadrilateral, preferably one that has no special properties, and locate the midpoints of its sides. Connecting these midpoints in sequence will always present us with a parallelogram. We show several of these awkward-shaped quadrilaterals generating such parallelograms in Figure 4.1. Some might be special parallelograms, such as squares, rectangles, and rhombuses, and others just general parallelograms.

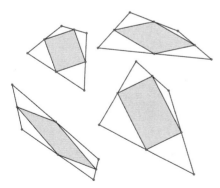

Figure 4.1.

After seeing this amazing phenomenon in geometry, you might ask what must be true about the original quadrilateral in order for the resulting parallelogram to be a square, a rectangle, or a rhombus?

To answer this question and to satisfy the reader's curiosity, we will make an exception to our usual method of presentation and give the reason why these parallelograms are formed. Figure 4.2 shows a quadrilateral, *ABDC*, where the midpoints of the sides are noted by points *E*, *F*, *H*, and *G*. We first focus on triangle *ABC*. Recall that when a line is drawn joining the midpoints of two sides of a triangle, it is one-half the length of and parallel to the third side. Therefore, *EF* is parallel to *BC*. Similarly, with triangle *BCD* we have *GH* parallel to *BC* and one-half of its length. Therefore, *EF* and *GH* are equal and parallel, which determines a parallelogram. Now let's take this a step further. Since the diagonals of quadrilateral *ABDC* are perpendicular, the sides of parallelogram *EFGH* also are perpendicular. This provides us with a parallelogram that is a rectangle.

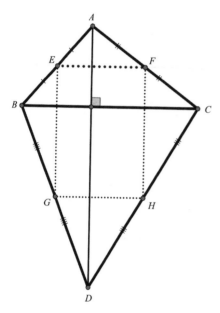

Figure 4.2.

If the diagonals are perpendicular and equal in length, as shown in Figure 4.3, the sides of the rectangle also are of equal length. The figure that results when the midpoints of the sides are joined is a square.

Continuing with this line of reasoning, suppose the diagonals are of the same length as those in Figure 4.4, where *AD* = *BC*. Then the lines joining the midpoints, each of which is one-half the length of a diagonal, must all be the same length. This results in a parallelogram that is a rhombus.

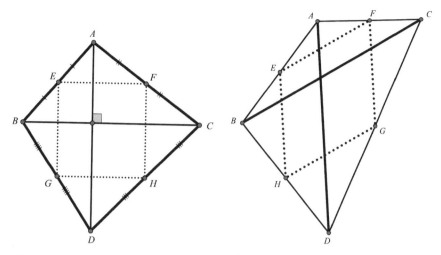

Figure 4.3.   **Figure 4.4.**

Note also that the perimeter of each parallelogram formed by joining the midpoints consecutively of a quadrilateral will always equal the sum of the lengths of the two diagonals. In addition, the area of the parallelogram formed by joining the midpoints of any quadrilateral is one-half the area of the original quadrilateral. These are truly geometric findings that should be appreciated.

Just to take this a step further, a parallelogram can also be formed by joining the midpoints of the diagonals with the midpoints of one pair of opposite sides of a quadrilateral. This is shown in Figure 4.5.

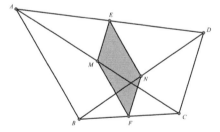

Figure 4.5.

Sometimes a rather strange way of creating a common geometric figure demonstrates the hidden beauty of geometry. Our next example shows how a rectangle can appear from very unexpected constructions. We begin with the cyclic quadrilateral *ABCD* inscribed in circle *O*, shown in Figure 4.6. We then construct the bisectors of the angles of the quadrilateral to meet the circumcircle at points *D*, *E*, *F*, and *G*. Unexpectedly, a rectangle appears as quadrilateral *EFGH*.

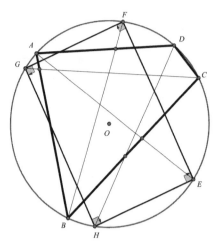

Figure 4.6.

A truly remarkable coincidence occurs when bisectors of opposite angles of a general quadrilateral intersect each other on a point on one of the diagonals. Unexpectedly, the bisectors of the other pair of opposite angles also meet, but on the other diagonal. This is illustrated in Figure 4.7, where the bisectors of angles B and D meet at point Q, which is on diagonal AC. When we draw the bisectors of angles A and C, we find that they meet at point P, which is on the other diagonal, BD. This is another example of the beautiful consistency of geometry that is often overlooked.

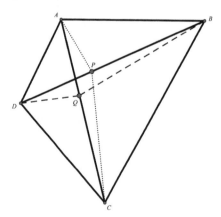

Figure 4.7.

Quadrilaterals continue to provide some quite unusual curiosities. Take, for example, quadrilateral $ABCD$ as shown in Figure 4.8, where diagonal BD divides the quadrilateral into two equal-area triangles, ABD and CBD.

When that is the case, we will always find a diagonal *BD* that divides diagonal *AC* into two equal parts, namely, *AP* = *PC*.

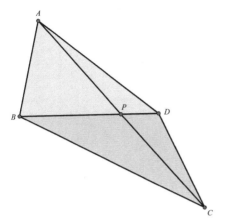

Figure 4.8.

Still working with a general quadrilateral, we find an unexpected relationship that can evolve when we construct the four bisectors of the angles of the quadrilateral. In Figure 4.9, we have drawn the bisectors, *AE*, *BG*, *CG*, and *DE*, of angles *A*, *B*, *C*, and *D*, respectively. You know that any three noncollinear points always determine a unique circle; however, it is not particularly common for four points to lie on the same circle. Yet, with this configuration the points of intersection, *E*, *F*, *G*, and *J*, of the adjacent angle bisectors all lie on the same circle, resulting in a cyclic quadrilateral *EFGJ*. This is quite noteworthy!

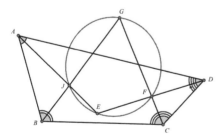

Figure 4.9.

There is one relationship of particular interest in the configuration shown in Figure 4.9. If the original quadrilateral *ABCD* is a parallelogram, then the resulting quadrilateral *EFGJ* will be a rectangle, as shown in Figure 4.10.

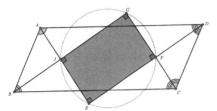

Figure 4.10.

Taking this one step further, if the original quadrilateral *ABCD* is a rectangle, then the resulting figure formed by the four angle bisectors of the rectangle will be a square. We can see this in Figure 4.11, where *EFGJ* is a square.

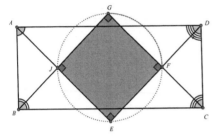

Figure 4.11.

This could be generalized to any cyclic quadrilateral. Suppose we take the angle bisectors and find their points of intersection on the circumcircle. Lo and behold, a rectangle is formed, and again the intersection points of the adjacent angle bisectors are concyclic. We show this in Figure 4.12, where the angle bisectors of quadrilateral *ABCD* determine the cyclic quadrilateral *PQRS*. When those bisectors meet the circumcircle of quadrilateral *ABCD*, those intersection points form rectangle *HEFG*.

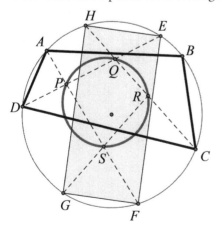

Figure 4.12.

We can take this a step further. When the diagonals of a cyclic quadrilateral are perpendicular, the resulting quadrilateral formed by the angle bisectors creates a square. This is shown in Figure 4.13, where diagonals *AC* and *BD* are perpendicular, and the points at which the bisectors of the angles of quadrilateral *ABCD* meet the circle create square *HEFG*.

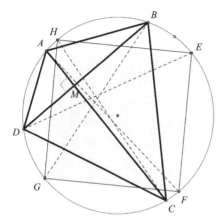

Figure 4.13.

Although we have dealt with general quadrilaterals, there is an extended version called a *complete quadrilateral*, which is formed by extending opposite sides to meet (assuming opposite sides are not parallel). This is shown in Figure 4.14, where *ABCDEAF* is a complete quadrilateral. A complete quadrilateral has three diagonals, *AD*, *CF*, and *BE*. The striking part about these diagonals is that their midpoints, *M*, *N*, and *K*, respectively, will always end up being collinear, as shown in Figure 4.14. Quite amazing!

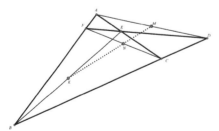

Figure 4.14.

In this particular situation, we have the original simple quadrilateral *FBCE* inscribed in a circle, as shown in Figure 4.15. Yet by extending the opposite sides and considering their points of intersection, we will have a complete quadrilateral. When we draw the bisectors of the opposite angles, *BAC* and *BDF*, the result is truly unexpected. The points of intersections with the opposite sides determine a rhombus, *GHJK*.

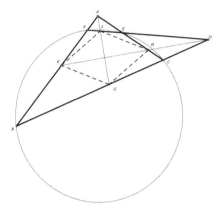

Figure 4.15.

An interesting feature of a cyclic quadrilateral, such as the one shown in Figure 4.16, is that the perpendicular bisectors of the sides of each of the four triangles, *ABC*, *CDA*, *BCD*, and *BAD*, are concurrent at the center of the circumcircle of the cyclic quadrilateral *ABCD*. Put another way, the perpendicular bisectors of each of the sides and the diagonals of the cyclic quadrilateral will all be concurrent at the center of the circumcircle.

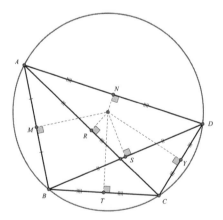

Figure 4.16.

Occasionally rather unusual geometric configurations lead to determining four points on one circle. One such arrangement is shown in Figure 4.17, where we begin with triangle *ABC* and line *PQ* parallel to side *BC*. We then construct a circle tangent to side *AC* and intersecting side *AB* at points *P* and *R*. Unexpectedly, we find that points *Q, R, B,* and *C* all lie on the same circle.

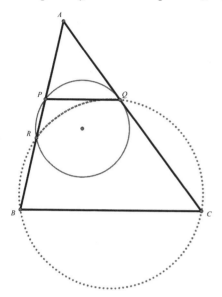

Figure 4.17.

Let us now consider a randomly drawn cyclic quadrilateral, where we construct a line from the midpoint of each side of the quadrilateral and perpendicular to the opposite side. In Figure 4.18, perpendiculars are drawn from the midpoints *E, F, G,* and *H* of sides *AD, AB, BC,* and *CD,* respectively, and perpendicular to the opposite sides of the quadrilateral. Quite unexpectedly, these four lines are all concurrent at point *P*.

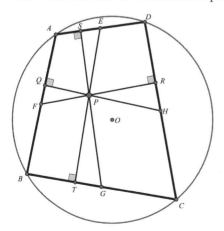

Figure 4.18.

Point *P* in Figure 4.18 has another coincidental property. In Figure 4.19, we once again have a cyclic quadrilateral and point *P* determined, but this time we draw the diagonals of quadrilateral *ABCD* and also locate the midpoints, *M* and *N*, of those diagonals. In triangle *KMN*, where *K* is the intersection of the two diagonals, we find that point *P* is also the orthocenter of triangle *KMN*.

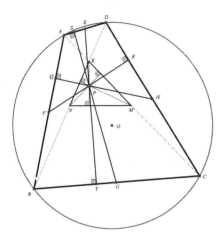

Figure 4.19.

Once again, point *P*, the intersection of the segments joining the mid-points of one side and perpendicular to the opposite side of a cyclic quadrilateral, plays an unexpected role. In Figure 4.20, when we extend a pair of opposite sides of the cyclic quadrilateral—in this case, sides *AD* and *BC*—they meet at point *X*. We then consider the line *EG* joining the midpoints of the same pair of opposite sides, which we extended. Unexpectedly, when we draw the perpendicular from point *X* to line *EG*, it contains a point, *P*, and thereby joins in the concurrency.

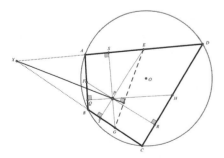

Figure 4.20.

This time let's consider a special cyclic quadrilateral that has perpendicular diagonals as shown in Figure 4.21, where *AC* is perpendicular to *BD*. The unexpected result here is that, if we draw a line from the midpoint of one side of the quadrilateral and perpendicular to the opposite side, this line is concurrent with the two perpendicular diagonals of the cyclic quadrilateral. In Figure 4.21, line *EF* joins the midpoint of *DC*, is perpendicular to *AB*, and turns out to be concurrent with the two diagonals, *AC* and *BD*, at point *P*.

Conversely, we can say that if a cyclic quadrilateral has perpendicular diagonals, then the perpendicular to one side of the quadrilateral from the point of intersection of the diagonals bisects the opposite side of the quadrilateral.

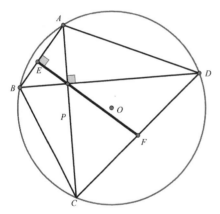

Figure 4.21.

Another unusual relationship evolves when considering cyclic quadrilaterals with perpendicular diagonals. If we draw a perpendicular from the center of the circle to one side of the quadrilateral, the length of the segment is half that of the opposite side of the quadrilateral. In Figure 4.22, *ON* is perpendicular to side *CD*. It then turns out that $ON = \frac{1}{2}AB$. Of course, this line could have been drawn from the center of the circle to any of the sides of the quadrilateral, and that perpendicular segment would be half the length of the opposite side. Think about it; quite astonishing!

A cyclic quadrilateral leads us to more unexpected relationships. Let's once again consider cyclic quadrilateral *ABCD*, shown in Figure 4.23. This time we draw the diameter of the circumscribed circle from point *A* to meet the opposite side of the circle at point *N*. It turns out that point *N* enables us to conclude that *BN* = *CD*.

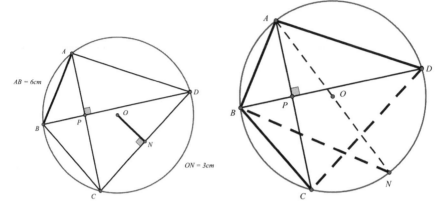

Figure 4.22.   **Figure 4.23.**

Midpoints of lines sometimes can produce unexpected results. For example, if we join the midpoints of the two diagonals of any cyclic quadrilateral, we find that this line is concurrent with the two lines joining the midpoints of the opposite sides of the quadrilateral. This is shown in Figure 4.24, where the midpoints of the two diagonals are points M and N, and the midpoints of the sides are E, F, G, and H. Point P is the intersection of the two lines joining the midpoints of the opposite sides, EG and FH, and it is also the midpoint of line MN. This situation can be seen as a concurrency of the three lines, EG, FH, and MN, or as a collinearity of points M, P, and N. There is still one more thing to marvel about in this configuration: point P is the midpoint of MN. The beauty of this arrangement lies in the fact that it is true for *any* cyclic quadrilateral.

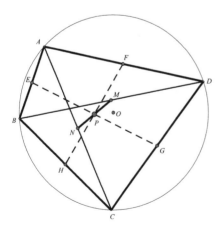

Figure 4.24.

When we combine two quadrilaterals, lots of interesting relationships can result. Figure 4.25 shows a cyclic quadrilateral *ABCD* inscribed in circle *O*. At each of the vertices of quadrilateral *ABCD*, a tangent to the circumcircle is drawn, creating quadrilateral *HKLJ*, which is circumscribed about circle *O*. The first amazing thing we find here is that the diagonals of the two quadrilaterals are all concurrent at point *P*. We also find a collinearity when we look at the complete quadrangle *ABCD*, where points *F*, *G*, and *E* are collinear. An ambitious reader can find other concurrencies and collinearities in this rather rich geometric configuration.

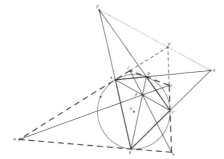

Figure 4.25.

One of the more famous relationships of cyclic quadrilaterals is a theorem attributed to Claudius Ptolemaeus of Alexandria (commonly referred to as Ptolemy). In his work, the *Almagest*, Ptolemy stated the following: the product of the lengths of the diagonals of a cyclic quadrilateral equals the sum of the products of the lengths of the pairs of opposite sides. Applying this theorem to Figure 4.26 yields:

$$AC \cdot BD = AB \cdot DC + AD \cdot BC.$$

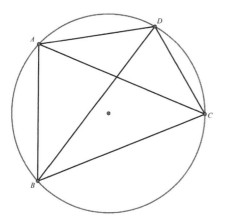

Figure 4.26.

This leads to some rather unusual length relationships. For example, suppose a parallelogram is intersected by a circle that contains one vertex and intersects two of its adjacent sides, which is shown in Figure 4.27. Here circle O passes through the vertex A of parallelogram $ABCD$ and intersects two sides and the diagonal at points P, Q, and R. When this is the case, the following strange relationship results:

$$AQ \cdot AC = AP \cdot AB + AR \cdot AD.$$

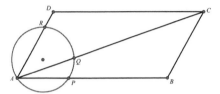

Figure 4.27.

Ptolemy's theorem provides some rather interesting relationships among the lengths of lines joining a point on the circumcircle of a regular polygon to each of the polygon's vertices. Here is a summary of the relationship of the first few regular polygons.

First, equilateral triangle ABC is inscribed in a circle with point P on the circle, as shown in Figure 4.28. The following is then true: $PA = PB + PC$.

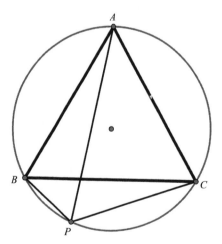

Figure 4.28.

The next regular polygon is a square, and in Figure 4.29 we show square *ABCD* with point *P* on the circumcircle. The following relationship results:
$$\frac{PD}{PA} = \frac{PA + PC}{PB + PD}.$$

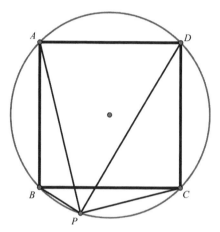

Figure 4.29.

Next is a regular pentagon, *ABCDE*, shown in Figure 4.30, with point *P* on the circumcircle. Here, Ptolemy's theorem provides us with the following:

$$PA + PD = PB + PC + PE.$$

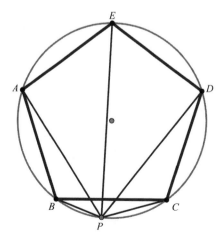

Figure 4.30.

Finally, we have the regular hexagon *ABCDEF*, pictured in Figure 4.31, and once again point *P* is on the circumcircle. This results in the following relationship:

$$PE + PF = PA + PB + PC + PD.$$

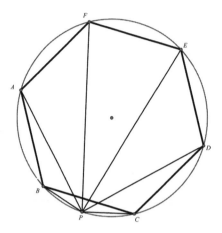

Figure 4.31.

$$5$$

On Circles

We have considerably immersed ourselves with circles to this point. But there are other beautiful relationships that exist primarily among circles—apart from those we have already discussed. One such relationship is based on the *arbelos*, pictured in Figure 5.1. Here the darker area bounded by three semicircles is mounted on one line, and the sum of the diameters of the two smaller semicircles is equal to the diameter of the larger semicircle.

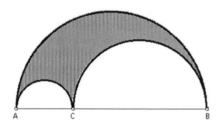

Figure 5.1.

There are lots of amazing features in this configuration, as shown in Figure 5.2. Here, GC is a common tangent to the two smaller semicircles, and SR is tangent to both smaller semicircles. The following are a few of these curiosities to appreciate—you may want to search for others!

$$\text{Arc } AGB = \text{Arc } ASC + \text{Arc } CRB$$

There are two sets of collinear points: A, S, and G, as well as B, R, and G. Lines SR and CG bisect each other at point P.
Points G, R, C, and S are concyclic with the center of the circle at point P.
The area of the arbelos is equal to the area of the circle with center P.

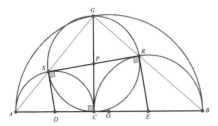

Figure 5.2.

Other curiosities are to be found in this arbelos configuration. For example, if we draw segments *RC* and *SC*, we unexpectedly end up with a rectangle, *RCSG*, as shown in Figure 5.3.

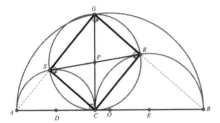

Figure 5.3.

We extend our appreciation of the arbelos even further by considering a circle tangent to each of the three semicircles, as shown in Figure 5.4. Many more such configurations exist, including other circles tangent to those in Figure 5.4.

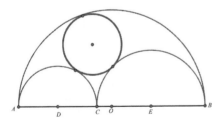

Figure 5.4.

There are almost always more fascinating things to find in any geometric configuration. Let's once again consider the arbelos. This time we draw a complete large circle and locate the midpoint of the semicircle, *M*, below the arbelos. We then create the odd-looking quadrilateral *SMRC*, shown in Figure 5.5. It can be demonstrated that the area of this odd-looking quadrilateral is equal to the sum of the squares of the radii of the two smaller semicircles. In equation form we would write this as follows: $Area\,SMRC = r_1^2 + r_2^2$.

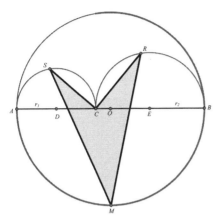

Figure 5.5.

Other analogous structures can provide geometric insights. For example, Figure 5.6 shows a configuration of two small equal semicircles and two larger semicircles encasing a region. Particularly fascinating here is that the area bounded by these four semicircles is equal to that of the large circle (shown in Figure 5.7) whose diameter is shown as the distance between the two larger semicircles, *AB*.

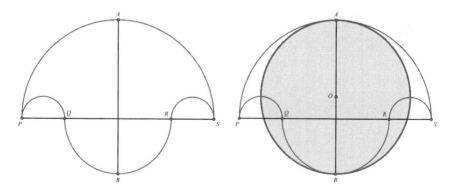

Figure 5.6.   **Figure 5.7.**

Several variations of this arrangement can be used for further entertainment and discoveries. Consider the configuration shown in Figure 5.8, where the sum of the diameters of the two smaller semicircles is equal to the diameter of the large semicircle. We draw a tangent from point *A* to the smaller semicircle at point *T*. Circle *O* is then drawn with *AT* as its diameter. The area of circle *O* turns out to be equal to that of the two smaller semicircles!

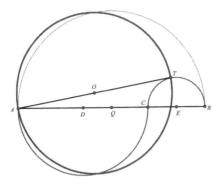

Figure 5.8.

Countless area comparisons and calculations can be performed with semicircles. Figure 5.9 shows one more such example, where the area mapped out by the bold curved lines (four semicircles) is equal to that of the complete circle shown with dashed lines. We can say that the area of *ABFCDE* is equal to that of the circle with center at *O*.

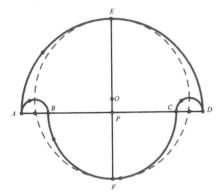

Figure 5.9.

These semicircle-generated figures lead to some interesting collinearities. Figure 5.10 shows a configuration of three semicircles centered at *M*, *C*, and *D*. When we draw a circular arc centered at *A* and tangent at point *E* to the circle centered at *C*, we find that points *E*, *C*, and *A* are collinear. When we draw a circle centered at *A* and tangent at point *F* to the semicircle centered at point *D*, we find a concurrence with points *D*, *F*, and *A*.

Most of the geometric figures presented always have additional features. For example, from Figure 5.10 we can very easily create a regular pentagon, as shown in Figure 5.11. Here we use the intersection points of the last two circular arcs with the large circle to determine four of the vertices of the regular pentagon.

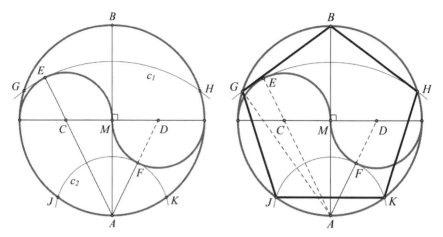

Figure 5.10.   **Figure 5.11.**

Perhaps one of the best recalled relationships in geometry is the Pythagorean theorem. It states that the sum of the squares *of* the legs of a right triangle is equal to the square *of* the hypotenuse. Restating this theorem by changing the word "of" to "on" gives it a geometric interpretation. Taking this a step further, we don't need *squares on* the hypotenuse and on the legs, but any similar figures would also suffice. For example, the sum of *the areas of* the semicircles on the legs of a right triangle is equal to *the area of* the semicircle on the hypotenuse. Thus, for Figure 5.12 we can say that the areas of the semicircles are related as follows: *Area P = Area Q + Area R.*

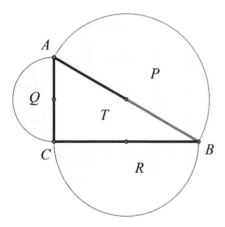

Figure 5.12.

Suppose we now flip semicircle *P* over the rest of the figure (using *AB* as its axis). We would get the configuration shown in Figure 5.13. Let us now focus on the *lunes*, L_1 and L_2, formed by the two semicircles.

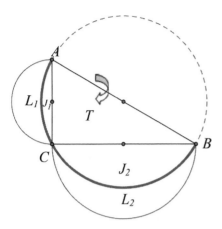

Figure 5.13.

The resulting diagram after we flip the semicircle will look like that in Figure 5.14. Earlier, in Figure 5.12, we established that *Area P = Area Q + Area R.* In Figure 5.14 that same relationship can be written as follows:

$$Area\ J_1 + Area\ J_2 + Area\ T = Area\ L_1 + Area\ J_1 + Area\ L_2 + Area\ J_2$$

If we subtract *Area J_1 + Area J_2* from both sides, we get an astonishing result:

$$Area\ T = Area\ L_1 + Area\ L_2$$

That is, we have a rectilinear figure (the triangle) equal to some nonrectilinear figures (the lunes). This is quite unusual, since the measures of circular figures seem to always involve π, while rectilinear (or straight-line) figures do not.

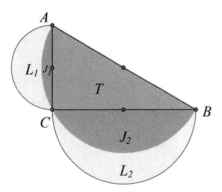

Figure 5.14.

An analogous situation can be gotten by extending the above scenario to that of a square, as shown in Figure 5.15. Here the sum of the areas of the four lunes is equal to the area of the square. In equation form we have: *Area ABCD = Area L$_1$ + Area L$_2$ + Area L$_3$ + Area L$_4$.*

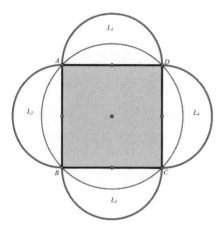

Figure 5.15.

Most of the results presented so far have been known for many centuries. The next "wonder," however, was first published in 1974 by John Evelyn, G. B. Money-Coutts, and J. A. Tyrrell in *The Seven Circles Theorem and Other New Theorems* (London: Stacey International, 1974). This implies that reasonably simple unknown results in elementary geometry are still out there, waiting to be discovered by some diligent researchers. Figure 5.16 shows a large circle, with six more circles packed inside the large circle. Each of these is tangent to the large circle at points P_1, P_2, P_3, P_4, P_5, and P_6, respectively, and any two successive circles among these also are tangent to each other. In other words, the circles through P_1 and P_2 touch in a point, as do those through P_2 and P_3, P_3 and P_4, P_4 and P_5, P_5 and P_6, as well as P_6 and P_1. If all of these pairs of circles are tangent, it follows that lines P_1P_4, P_2P_5, and P_3P_6 pass through a common point, Q. This is true under quite general circumstances.

We now go from the seven circles theorem to the famous five circles theorem. Figure 5.17 shows a central circle with five circles intersecting each other placed sequentially around it. When we consecutively connect the internal intersection points, we form a pentagram whose vertices lie on each of the five circles.

The ambitious reader may want to study other multiple circle theorems, which are readily available. Here we merely whet the reader's appetite to do further research.

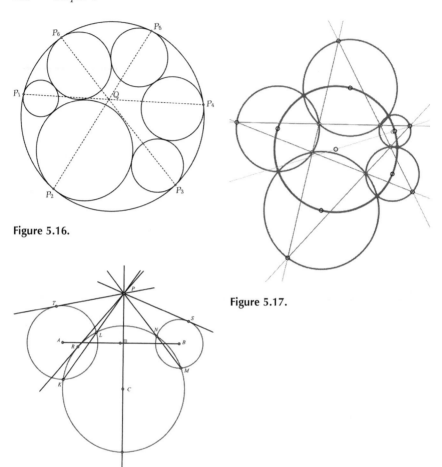

Figure 5.16.

Figure 5.17.

Figure 5.18.

Circles also can generate some concurrency, as shown in Figure 5.18. Here we have three circles whose centers are not collinear, and which intersect in pairs. We will join the pairs of intersection points and the tangents to each of the circles, noticing how they are all concurrent and of equal length. Put another way, when we first located point P, the intersection of KL and MN, we found that the tangents from P to each of the circles are the same length, that is, $PT = PR = PS$. (We chose only one tangent to each of the other three circles, since the other three tangents would clearly be the same length.)

6

Admiring Other Geometric Phenomena

Our next journey through geometry will present a variety of unstructured curiosities spanning a plethora of ideas. We begin rather gently by introducing some simple constructions and then delve into some beautiful and unexpected geometric relationships.

We start by showing how easy isosceles triangles are to construct. We take the given equal side lengths and make them the radius of the circle. Then we choose the desired measure for the vertex angle and draw the other radius at that angle measure. Figure 6.1 shows one such simple construction resulting in triangle *ABC*.

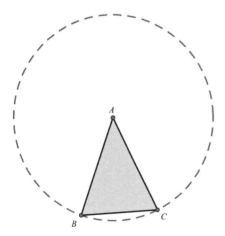

Figure 6.1.

Other rather unusual constructions result in an isosceles triangle. Consider isosceles triangle *ABC*, shown in Figure 6.2. We can select any point *P* along *BC* and erect a perpendicular line that would intersect the other two

101

sides (extended) at points D and E. Unexpectedly, triangle ADE will always be an isosceles triangle, with $AE = AD$.

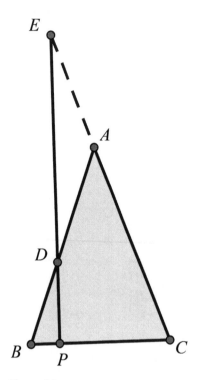

Figure 6.2.

Sometimes locating points on the circumcircle of a triangle can be surprising. Let's consider triangle ABC with altitudes AX, BY, and CZ, which determine the orthocenter P, as shown in Figure 6.3. We then choose any point D on side BC and draw a circle with center D and radius DP. When altitude AX is extended to meet this circle at point E, we find that point A is also on the circumcircle of the original triangle ABC. How curious it is for two circles to meet at a point determined independently of the circles—further evidence of the beauty of geometry.

Speaking of strange configurations that lead to unexpected results, let's consider triangle ABC shown in Figure 6.4, where AD is the bisector of angle BAC. Through point B we construct a line parallel to the angle bisector AD, which meets line CA extended at point H. When we construct the circumcircle of triangle ABC and the circle determined by points C, D, and H, we find that the two circles intersect line AD extended at points N and E, which just happen to be two points equidistant from point A, or put another way: $AN = AE$.

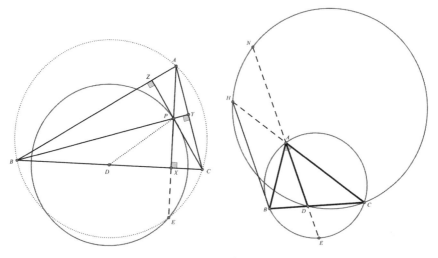

Figure 6.3.   **Figure 6.4.**

Here, we have another example of how mathematics produces an equality when it is least expected.

Sometimes just drawing a few circles also leads us to some equal line segments. Consider right triangle *ABC*, shown in Figure 6.5, where we draw a circle on each side of the right triangle, such that each of the sides is the diameter of the respective three circles. We then simply draw any line from point *A* to cut each of the three circles at points *F*, *H*, and *G*. Quite unexpectedly, we find that *AH* = *FG*. What makes this so remarkable is that it is true for any right triangle!

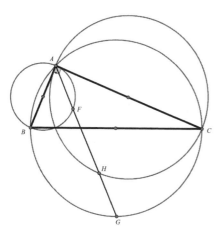

Figure 6.5.

Other geometric configurations also lead to parallel lines rather simply. We begin with triangle *ABC*, shown in Figure 6.6, with median *AM*. We draw two lines, each from one of the other two vertices of the triangle, so that they intersect the median, *AM*, as well as the opposite sides *AB* and *AC* at points *D* and *E*, respectively. The result is that line *DE* is parallel to *BC*.

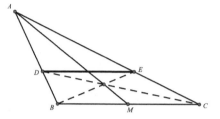

Figure 6.6.

Constructing a right angle is normally a rather simple procedure. Sometimes, however, we might want to draw a line parallel to the base of a triangle and at the same time create a right angle with the vertex at any given point on the base. This may seem a bit contrived, but it does once again demonstrate the hidden beauty in geometry. We begin with triangle *ABC*, shown in Figure 6.7. We seek the precise place at which line *FG* can be drawn parallel to the base, *BC*, so that the two points, *F* and *G*, at which it intersects the remaining two sides of the triangle will enable us to construct the right angle whose vertex is on point *P* on side *BC*.

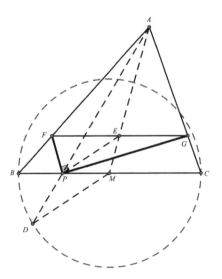

Figure 6.7.

We begin the construction by locating the midpoint, *M*, of side *BC* and drawing a circle with center *M* and radius *MC*. We then draw line *AP* to intersect the circle at point *D*. This allows us to draw line *DM*; whereupon we then draw a line containing point *P* and parallel to *DM*, intersecting line *AM* at point *E*. This allows us to construct a line through point *E* and parallel to *BC*, intersecting sides *AB* and *AC* of triangle *ABC* at points *F* and *G*, respectively. By drawing lines *FP* and *GP*, we will have created the right angle *FPG* at the predetermined point *P* on line *BC*, which was our initial goal. This is a rather difficult construction for a seemingly easy task, but it once again demonstrates geometry's power.

Sometimes, we seek to find the longest line that can be drawn within a given configuration such as that shown in Figure 6.8. Here, we have two semicircles of which the radius of one is the diameter of the other, and where we seek to find the longest line that would be perpendicular to the common radius/diameter line with its endpoints on each of the semicircles. The diagram shows that *C* is the midpoint of diameter *AB*, and *D* is the midpoint of diameter *BC*. Point *G*, at which *CD* is trisected, so that *CG* = 2*DG*, turns out to be the point at which *EF* is the longest perpendicular with endpoints on the two semicircles.

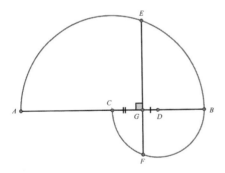

Figure 6.8.

We now embark on a rather strange path to construct a rhombus beginning with a random quadrilateral, yet with equal diagonals, as shown in Figure 6.9. First, we draw on each side of the original quadrilateral *ABCD* a circle with the side of the quadrilateral as its diameter.

Next, we just draw the four common chords, *ANL*, *BFL*, *DKJ*, *CEJ*, of each pair of circles. We will then have created rhombus *SJRL* (Figure 6.10). A strange and unexpected result!

Constructing an equilateral triangle using straightedge and compasses (or a dynamic geometry program such as *GeoGebra* or *Geometer's Sketchpad*) is a rather simple process. We merely choose the length of the side of the

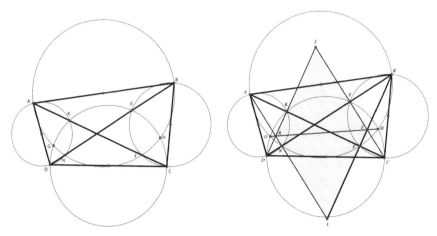

Figure 6.9.   **Figure 6.10.**

to-be-constructed equilateral triangle, as shown in Figure 6.11, and construct the circle with *B* as center and radius *BC* and then the circle with *C* as center and radius *BC*. The point at which the two circles intersect will be point *A*, resulting in an equilateral triangle, *ABC*.

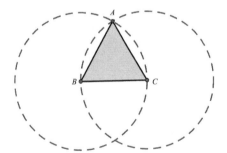

Figure 6.11.

Another surprisingly simple construction for an equilateral triangle begins with an isosceles triangle whose vertex angle is 120°. All we need to do is to locate the trisection points along the base of the isosceles triangle and connect them to the vertex, and we have an equilateral triangle. We show this in Figure 6.12, where ∠*BAC* = 120° and points *D* and *E* are the trisection points of the base, *BC*. The resulting triangle *ADE* is equilateral.

It would be interesting to see how to construct an equilateral triangle equal in area to that of a given triangle, although this is not often done. The process may look complicated, but just follow along and you will see how it makes sense.

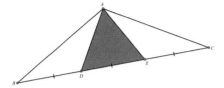

Figure 6.12.

We begin with the given triangle *ABC*, shown in Figure 6.13, for which we would like to construct an equilateral triangle with an equal area. We begin by constructing an equilateral triangle (something we have reviewed earlier) using *BC* as a side, thus creating equilateral triangle *DBC*. Next, through point *A*, we draw a line parallel to *BC* that intersects line *DB* at point *E*. At point *E*, we erect a perpendicular to line *DB*. After locating the midpoint of *DB* and creating a semicircle on diameter *DB*, we will call the point of intersection with the perpendicular we just created, point *F*. We draw a circular arc with center *B* and radius *BF*, meeting *DB* at point *G*. Through point *G*, we construct a parallel line to *DC*. This then gives us the required equilateral triangle *BGH*, which is equal in area to triangle *ABC*. Although a bit complicated, the mission is completed!

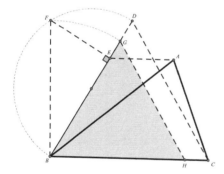

Figure 6.13.

In 1900, American mathematician Frank Morley (1860–1937) published a remarkable geometric relationship that can be applied to any shape triangle. It simply states that the angle trisectors of any triangle can determine an equilateral triangle. Figures 6.14, 6.15, 6.16, and 6.17 shows various triangles of different shapes, and in each case we have the trisectors of its angles. We mark the intersections of adjacent trisectors as points *D*, *E*, and *F*. In each case, the triangle formed by these three points is always an equilateral triangle. This is truly a remarkable theorem, which reminds us of our discoveries in Chapter 1 on concurrency, as you will see when we explore this wonderful finding.

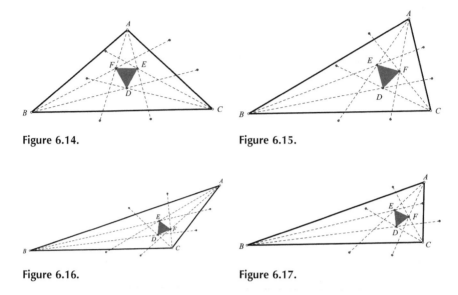

Figure 6.14.   **Figure 6.15.**

Figure 6.16.   **Figure 6.17.**

When we connect the vertices of the original triangle *ABC* with corresponding vertices of the equilateral triangle, *DEF*, formed by the trisectors, we find that they are concurrent. And so in Figure 6.18 we once again have a concurrency, thereby demonstrating the beauty and consistency in geometry.

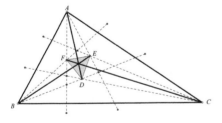

Figure 6.18.

One rather unusual arrangement leads to an unexpected equality. Suppose we have two triangles, *ABC* and *PBC* (Figure 6.19), whose only relationship is that they share the same base, *BC*, and their third vertex lies on a line parallel to *BC* (that is, *AP* is parallel to *BC*). We now choose any line parallel to *BC* and extend the sides of the two triangles to meet that parallel line at points *D*, *E*, *F*, and *G*. Regardless of the shapes of the two triangles and how far below the triangle the third parallel line is, you always have *DE* = *FG*. Amazing but true!

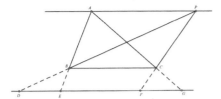

Figure 6.19.

If we consider a parallelogram with a point, P, anywhere in its interior, we can establish a very interesting area relationship. In Figure 6.20, point P is placed within parallelogram $ABCD$. From P we draw lines to each of the four vertices of the parallelogram. It turns out that $Area\triangle APB = Area\triangle DPC + Area\triangle APC + Area\triangle BPD$. What makes this so special is that the point P, which we selected to be *anywhere* within the parallelogram, and the two diagonals we use to form the triangles produced this truly unexpected result.

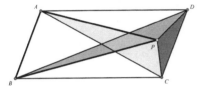

Figure 6.20.

Parallelograms often lend themselves to some unexpected properties. Take, for example, parallelogram $ABCD$ shown in Figure 6.21, where any two parallel lines, AF and EC, are selected within the parallelogram, and where points F and E are located on sides DC and AB, respectively. From point F we draw a line parallel to diagonal AC meeting side AD at point P. When we draw line segment PE, we find that it is parallel to the other diagonal BD. Remember, points F and E could have been anywhere along the sides of the parallelogram as long as they generated the two parallel lines AF and CE.

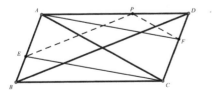

Figure 6.21.

And speaking of parallelograms, here is a simple-looking question that has stumped lots of people: What is the relationship between the two parallelograms, *ABCD* and *BEFP*, shown in Figure 6.22, where point *P* is on *AD* and point *C* is on *EF*? The two parallelograms share a common vertex point, *B*. In the attempt to find a solution, various lines are drawn and the ease of getting the answer is lost. Don't look ahead! Try to answer the question without looking further ahead.

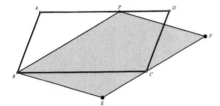

Figure 6.22.

All that is needed is to draw one line, *PC*, as we have done in Figure 6.23. We then notice that triangle *BPC* is one-half the area of parallelogram *ABCD*, since it shares base *BC* and has the same altitude from *P* to base *BC*. Analogously, triangle *BPC* is also one-half the area of parallelogram *BEFP*, since it shares base *BP* and has the same altitude from base *BP* as has parallelogram *BEFP*. Therefore, since triangle *BPC* is one-half the area of each of the parallelograms, the parallelograms must be equal in area.

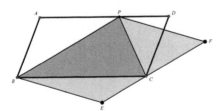

Figure 6.23.

Here is further evidence that parallel lines can evolve when they are least expected. In Figure 6.24 we begin with triangle *ABC* and mark the midpoints of its sides at points *D, E,* and *F*. We select any point on *EF* and call it point *G*. We then draw a line from *A* through *G* to meet *DE* at point *H*. Oddly enough, lines *GC* and *BH* end up being parallel.

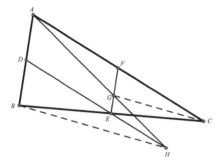

Figure 6.24.

Here we create two equal angles in a situation where they would least be expected to arise. In Figure 6.25, right triangle *ABC*, with a right angle at vertex *A* and the altitude from *A*, intersects the hypotenuse *BC* at point *D*. From point *D* perpendiculars are drawn to the other two sides of the triangle, meeting them at points *M* and *N*. We end up creating equal angles *BMC* and *BNC* by drawing lines *MC* and *NB*. It is surprising how equal angles can emerge in any right triangle that follows this procedure.

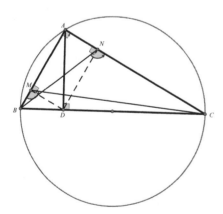

Figure 6.25.

Sometimes a very complicated-looking figure ends up unexpectedly yielding line segments of equal length. That is the case in Figure 6.26, where right triangle *ABC* is inscribed in circle *O*, and *D* is *any point* on arc *AC*. From point *D* a perpendicular is drawn to diameter *CB*, meeting it at point *E* and intersecting *AC* at point *F*. Lastly, a perpendicular is drawn to *AC* at point *F* and intersecting the circle determined by diameter *AC* at point *G*. The result of all these constructions is that three equal line segments, *GC*, *DC*, and *JC*, appear.

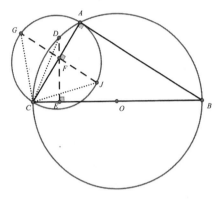

Figure 6.26.

7

The Golden Rectangle

For centuries, artists and architects have identified what they believed to be the ideal rectangle. This rectangle, often referred to as the *golden rectangle*, has also proved to be the most pleasing to the eye. The golden rectangle has the following ratio of length and width: $\phi = \dfrac{w}{l} = \dfrac{l}{w+l}$. This is known as the *golden ratio*, symbolized by the letter ϕ. (See A. S. Posamentier and I. Lehmann, *The Glorious Golden Ratio* [Amherst, New York: Prometheus Books, 2012].)

The desirability of this rectangle has been borne out by numerous psychological experiments. For example, Gustav Theodor Fechner (1801–1887), a German experimental psychologist, inspired by German philosopher Adolf Zeising's (1820–1876) book *Der goldene Schnitt* (The Golden Section),[1] began to investigate whether the golden rectangle had a special psychological aesthetic appeal in *Neue Lehre von den Proportionen des menschlichen Körpers* (New Theories about the Proportions of the Human Body)[2]. His findings were published in 1876 in *Zur experimentalen Ästhetik* (On Experimental Aesthetics)[3]. Fechner made thousands of measurements of commonly seen rectangles, such as playing cards, writing pads, books, windows, and other objects. He found that most had a ratio of length to width that was close to ϕ. He also tested people's preferences and found that most people preferred the shape of the golden rectangle.

In his research Fechner asked 228 men and 119 women which rectangle was aesthetically the most pleasing. Looking at the rectangles shown in Figure 7.1, which would you choose as the most pleasing to look at?

We can easily eliminate rectangle 1:1, as a square is considered by the general public not to be representative of a rectangle. It is, after all, a square!

---

[1] Published posthumously in 1884 by the Leopoldinisch Carolinische Akademie, Halle, Germany.
[2] Published in 1854 by R. Weigel, Leipzig.
[3] Published by Breitkopf & Härtel, Leipzig.

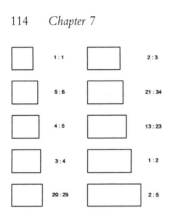

Figure 7.1.

Rectangle 2:5 (the other extreme) is uncomfortable to look at since it requires the eye to scan it horizontally. Rectangle 21:34, on the other hand, can be appreciated at a single glance. Fechner's findings seem to bear this out. The results that Fechner reported are shown in Table 7.1.

Fechner's experiment has been repeated with variations in methodology many times, and his results have been further supported. For example, in 1917 Edward Lee Thorndike (1874–1949), the American psychologist and educator, carried out similar experiments, with analogous results. In general, the rectangle with the ratio of 21:34 was most preferred. These two numbers are part of a Fibonacci sequence, 1, 1, 2, 3, 5, 8, 13, 21, 34, 55, 89, 144, …, where the ratio of consecutive numbers approaches the golden ratio (see A. S. Posamentier and I. Lehmann, *The Fabulous Fibonacci Numbers* [Amherst, New York: Prometheus Books, 2007]). Thus, $\frac{21}{34} = 0.\overline{61764705882352941} \approx \phi$, and hence, it creates the

Table 7.1.   Gustav Fechner's survey results

Ratio of sides of rectangle	Percent response for best rectangle	Percent response for worst rectangle
1:1 = 1.00000	3.0	27.8
5:6 = .83333	.02	19.7
4:5 = .80000	2.0	9.4
3:4 = .75000	2.5	2.5
20:29 = .68966	7.7	1.2
2:3 = .66667	20.6	0.4
21:34 = .61765	**35.0**	**0.0**
13:23 = .56522	20.0	0.8
1:2 = .50000	7.5	2.5
2:5 = .40000	1.5	35.7
	100.00	100.00

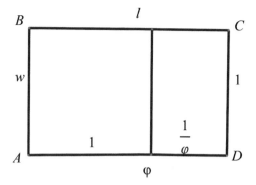

Figure 7.2.

golden rectangle, where the length, l, and the width, w, are in the following proportion: $\dfrac{w}{l} = \dfrac{l}{w+l} = \phi$. (See Figure 7.2.)

By multiplying means and extremes of this proportion we get $w^2 + wl = l^2$ or $w^2 + wl - l^2 = 0$. If we let $l = 1$, then $w^2 + w - 1 = 0$. Using the quadratic formula, we get $w = \dfrac{-1 \pm \sqrt{5}}{2}$. Because we have lengths, we use only the positive value. Therefore, $w = \dfrac{-1+\sqrt{5}}{2} = \dfrac{\sqrt{5}-1}{2} = \dfrac{1}{\phi}$ and $\phi = \dfrac{\sqrt{5}+1}{2}$.

Let's see how this rectangle may be constructed using the traditional Euclidean tools (an unmarked straightedge and compasses) or a computer program such as *Geometer's Sketchpad* or *GeoGebra*. With a width of 1 unit, our objective is to get the length to be $\dfrac{1+\sqrt{5}}{2}$ so that the ratio of the length to the width will be ϕ.

One of the simpler ways to construct this golden rectangle is to begin with a square, $ABEF$, shown in Figure 7.3, where M is the midpoint of AF. Then with

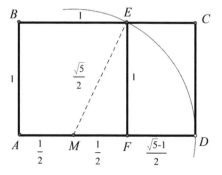

Figure 7.3.

radius *ME* and center *M*, draw a circle to intersect line *AF* at *D*. The perpendicular at *D* intersects line *BE* at *C*. We now have *ABCD*, a golden rectangle.

Let us continue with golden rectangle *ABCD*, where a square is constructed internally (as shown in Figure 7.4). If *AF* = 1 and *AD* = ϕ, then $FD = \phi - 1 = \frac{1}{\phi}$. We can establish that rectangle *CDFE* has dimensions $FD = \frac{1}{\phi}$ and *CD* = 1. If we inspect the ratio of length to width of rectangle *CDFE*, we get $\frac{EF}{FD} = \frac{1}{\frac{1}{\phi}} = \phi$. It is, therefore, also a golden rectangle.

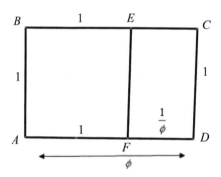

Figure 7.4.

We continue this process by constructing an internal square in the newly formed golden rectangle. In golden rectangle *CDFE*, square *DFGH* is constructed as shown in Figure 7.5. We find that $CH = 1 - \frac{1}{\phi} = \frac{1}{\phi^2}$, so the ratio of the length to the width of rectangle *CHGE* is $\frac{\frac{1}{\phi}}{\frac{1}{\phi^2}} = \phi$ (after multiplying both

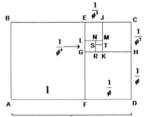

Figure 7.5.

numerator and denominator by ϕ^2). This, therefore, establishes that rectangle *CHGE* also is a golden rectangle.

Continuing this process, we construct square *CHKJ* in golden rectangle *CHGE*. We know that $\phi - \dfrac{1}{\phi} = 1$, therefore, $\phi - 1 = \dfrac{1}{\phi}$ and

$$EJ = \frac{1}{\phi} - \frac{1}{\phi^2} = \frac{\phi - 1}{\phi^2} = \frac{\dfrac{1}{\phi}}{\phi^2} = \frac{1}{\phi^3}.$$ We now inspect the ratio of the dimensions of

rectangle *EJKG*. This time, the length-to-width ratio is $\dfrac{\dfrac{1}{\phi^2}}{\dfrac{1}{\phi^3}} = \phi$. Once again,

we have a new golden rectangle, which is rectangle *EJKG*. By continuing this procedure, we get golden rectangles *GKML*, *NMKR*, *MNST*, and so on. Suppose we now draw the following quarter circles:

center *E*, radius *EB*
center *G*, radius *GF*
center *K*, radius *KH*
center *M*, radius *MJ*
center *N*, radius *NL*
center *S*, radius *SR*
etc.

The drawing, as shown in Figure 7.6, winds up being an approximation of a logarithmic spiral. The symmetric parts of this complex-looking figure are the squares. Suppose we locate the center of each of these squares. If we draw arcs through each of these points, we see that the centers of these squares lie in another approximation of a logarithmic spiral. This configuration is shown in Figure 7.7.

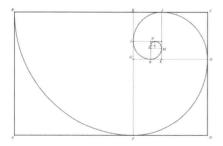

Figure 7.6.

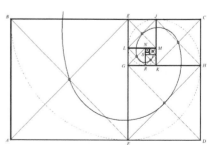

Figure 7.7.

The spiral in Figure 7.8 seems to converge (i.e., end) at a point in rectangle *ABCD*. This point is at the intersection, *P*, of *AC* and *ED*, which we can better see in Figure 7.8. Consider once again golden rectangle *ABCD*. Earlier we established that square *ABEF* determined another golden rectangle, *CEFD*. In Figure 7.8, we see that rectangles *ABCD* and *CEFD* are reciprocal rectangles. Furthermore, we see that reciprocal rectangles have corresponding diagonals that are perpendicular.

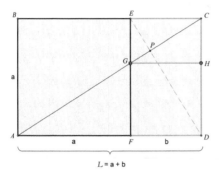

Figure 7.8.

In the same way as before, we can establish that rectangles *CEFD* and *CEGH* are reciprocal rectangles. Their diagonals, *ED* and *CG*, are perpendicular at *P*. This may be extended to each pair of consecutive golden rectangles shown in Figure 7.9. Clearly *P* ought to be the limiting point of the spiral.

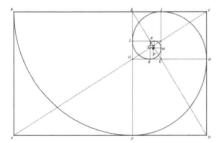

Figure 7.9.

We can use this relationship of the diagonals to construct consecutive golden rectangles. We could simply construct a perpendicular from *D* to *AC* in golden rectangle *ABCD*, and from its intersection *E* with *BC* construct a perpendicular to *AD* to complete the second golden rectangle. This process can be repeated indefinitely.

THE DIAGONALS OF THE GOLDEN RECTANGLE

We have done quite a bit with the golden rectangle, yet there never seems to be a limit to what you can do. For example, the golden rectangle— whose length and width are in the golden ratio—provides us a neat way to find the point along the diagonal that cuts it into the golden ratio. It is the unique properties of this special rectangle that enable us to do this so easily.

Consider the golden rectangle $ABCD$, whose sides $AB = a$ and $BC = b$, so that $\frac{a}{b} = \phi$. As shown in Figure 7.10, two semicircles are drawn on sides AB and BC to intersect at point S. If we now draw line segments SA, SB, and SC, we find that angles ASB and BSC are right angles (since they are each inscribed in a semicircle). Therefore, AC is a straight line, namely the diagonal. We now have the unexpected result that point S divides the diagonal in the golden ratio.

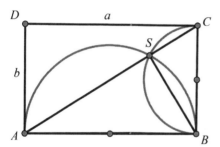

Figure 7.10.

The golden ratio can be created in practically endless ways. Consider the semicircle with the three congruent circles inscribed so that the tangency points are as shown in Figure 7.11.

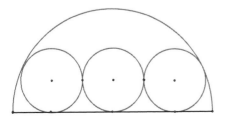

Figure 7.11.

We seek to find the ratio of the radius of the large semicircle to that of one of the smaller circles. In Figure 7.12, $AB = 2R$ and $AM = R$. Each of the congruent small circles has a radius r. Consider right triangle CEM with legs r and $2r$, where the hypotenuse then has length $r\sqrt{5}$. We now have $ME = r\sqrt{5}$ and $KE = r$, therefore, $MK = r(\sqrt{5}+1) = R$. Put another way, $\dfrac{R}{r} = \sqrt{5}+1 = 2\phi$. In this seemingly unrelated configuration of three congruent circles inscribed in a semicircle, we find the ratio of their radii is related to the golden ratio.

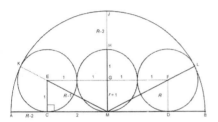

Figure 7.12.

THE GOLDEN TRIANGLE

We have thoroughly investigated the famous golden rectangle. Now we are ready to consider the golden ratio as it pertains to the golden triangle. As you would expect, the golden triangle, much like the golden rectangle, has Fibonacci numbers embedded within it, so it too consequently exhibits the golden ratio. Let's consider a triangle that contains the golden ratio. We begin by placing an isosceles triangle into another similar isosceles triangle, rather like we embedded our similar golden rectangles earlier. To do this, we draw the configuration shown in Figure 7.13. The sum of the angles of triangle ABC is $a + a + a + 2a = 5a = 180°$, and $a = 36°$.

This clearly leads us to a triangle with the angle measurements shown in Figure 7.14. Simple calculations show us that the ratio $\dfrac{\text{side}}{\text{base}} = \dfrac{1}{x} = \phi$ in triangle ABC.

We therefore call this a *golden triangle*. One easy way to construct a golden triangle is to first construct the golden section (done earlier in this chapter). Then draw a circle O with the longer segment of the golden section as radius OB, as shown in Figure 7.15. Then draw a circle, A, with the smaller segment AB of the golden ratio as the radius, centered at any point on the larger circle. The intersection point of the two circles, as shown in Figure 7.15, determines a golden triangle.

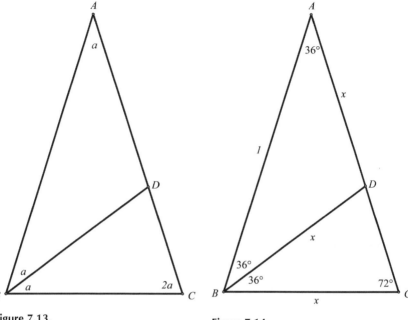

Figure 7.13.

Figure 7.14.

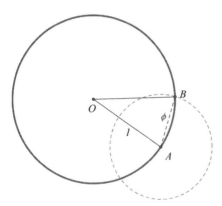

Figure 7.15.

By taking consecutively the following angle bisectors, *BD*, *CE*, *DF*, *EG*, and *FH*, of a base angle of each newly formed 36°, 72°, 72° triangle, we get a series of golden triangles (see Figure 7.16). These golden triangles (36°, 72°, 72°) are triangles: *ABC*, *BCD*, *CDE*, *DEF*, *EFG*, and *FGH*. Obviously, had space permitted we could have continued to draw angle bisectors and thereby generate more golden triangles. Our inspection of the golden triangle will be analogous to that of the golden rectangle.

Let us begin by having $HG = 1$ (Figure 7.16). Since the ratio $\frac{\text{side}}{\text{base}}$ of a golden triangle is ϕ, we find the following:

For golden triangle FGH: $\frac{GF}{HG} = \frac{\phi}{1}$, or $\frac{GF}{1} = \frac{\phi}{1}$, and $GF = \phi$.

Similarly, for golden triangle EFG: $\frac{FE}{GF} = \frac{\phi}{1}$, but $GF = \phi$, so $FE = \phi^2$.

In golden triangle DEF: $\frac{ED}{FE} = \frac{\phi}{1}$, but $FE = \phi^2$, therefore $ED = \phi^3$.

Again, for triangle CDE: $\frac{DC}{ED} = \frac{\phi}{1}$, but $ED = \phi^3$, therefore $DC = \phi^4$.

For triangle BCD: $\frac{CB}{DC} = \frac{\phi}{1}$, but $DC = \phi^4$, therefore $CB = \phi^5$.

Finally, for triangle ABC: $\frac{BA}{CB} = \frac{\phi}{1}$, but $CB = \phi^5$, therefore $BA = \phi^6$.

So, we see that the golden ratio is embedded throughout the figure.

As we did with the golden rectangle, we can generate an approximation of a logarithmic spiral by drawing arcs to join the vertex angle vertices of

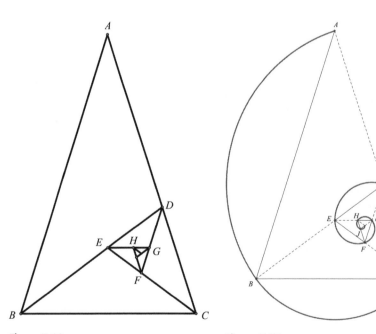

Figure 7.16.   **Figure 7.17.**

consecutive golden triangles (see Figure 7.17). We draw the circular arcs as follows: begin with circular arc *AB* centered at point *D*; then draw circular arcs *BC* centered at point *E*, *CD* centered at point *F*, *DE* centered at point *G*, *EF* centered at point *H*, and *FG* centered at point *J*. And so the spiral is created.

Many other truly fascinating relationships emanate from the golden ratio. Now that you have been exposed to the golden triangle, we next turn to the regular pentagon and regular pentagram (the five-pointed star) for more applications, since these are essentially composed of many golden triangles. You will then see that the golden ratio abounds throughout these shapes.

THE PENTAGON AND THE PENTAGRAM

A beautiful geometric shape, the regular pentagram, which was the symbol of the Pythagoreans, sums up much of the golden ratio in one configuration. The golden triangle is embedded many times in this shape (see Figures 7.18 and 7.19). According to Pythagoras, all geometric shapes can be described in terms of integers. So he was greatly disappointmented when one of his followers, Hippasus of Metapontum (ca. 450 BCE), showed that the ratio of the regular pentagon's diagonal to its side length could not be expressed as a fraction with integers. In other words, this ratio is not rational! This characteristic carried over to the Pythagoreans' symbol, the pentagram. The secret society was a bit troubled by this anomaly—which today can be seen as the beginning of our concept of irrational numbers (i.e., numbers that cannot be expressed as a ratio of two whole numbers, hence the name ir*rational*). In the regular pentagon, the ratio of the diagonal to the side is irrational. But which irrational number did Hippasus find? You guessed it! It was the golden ratio, ϕ.

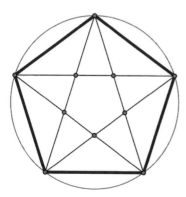

Figure 7.18.

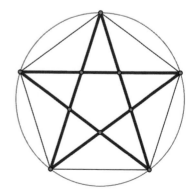

Figure 7.19.

To show that this length relationship is irrational, we use the fact that, in a regular pentagon, every diagonal is parallel to the sides it does not intersect. In triangles *AED* and *BTC* have parallel sides, so they are similar to each other.

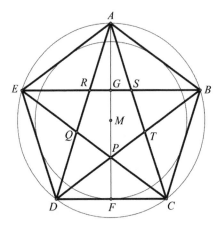

Figure 7.20.

Therefore $\dfrac{AD}{AE} = \dfrac{BC}{BT}$. But $BT = BD - TD = BD - AE$. In the regular pentagon, therefore, the following ratio holds: $\dfrac{\text{diagonal}}{\text{side}} = \dfrac{\text{side}}{\text{diagonal} - \text{side}}$.

In equation form we can write this as $\dfrac{d}{s} = \dfrac{s}{d-s}$ or $\dfrac{d}{s} = \dfrac{1}{\dfrac{d}{s} - 1}$ (with d as the length of the diagonal, and s the length of the side).

If we now let $x = \dfrac{d}{s}$ we get the equation $x = \dfrac{1}{x-1}$. This equation can be converted to the quadratic equation $x^2 - x - 1 = 0$, of which $\dfrac{d}{s}$ is a positive root, which just happens to be the irrational number $\phi = \dfrac{\sqrt{5}+1}{2}$. (Remember: $\sqrt{5}$ is irrational!)

This is what we claimed at the outset: the ratio of the diagonal to the side of a regular pentagon is irrational. As the irrational number $\pi = 3.1415926535897932384\ldots$ is inseparably connected with the circle, so too the irrational number $\phi = 1.6180339887498948482\ldots$ is connected inseparably with the regular pentagon!

The regular pentagon is a fascinating figure with lots of useful properties. We now present some for you to appreciate and ponder over. You might look for other such properties. In Figure 7.21 the regular pentagon *ABCDE* has the following properties:

1. The size of every interior angle is 108°: ∠*EAB* = ∠*ABC* = ∠*BCD* = ∠*CDE* = ∠*DEA* = 108°
   - ∠*BEA* = ∠*CAB* = ∠*DBC* = ∠*ECD* = ∠*ADE* = 36°,
   - ∠*PEB* = ∠*QAC* = ∠*RBD* = ∠*SCE* = ∠*TDA* = 36°,
   - ∠*CDA* = ∠*DEB* = ∠*EAC* = ∠*ABD* = ∠*BCE* = 72°.

2. Triangles Δ*DAC*, Δ*EBD*, Δ*ACE*, Δ*BDA*, Δ*CEA*, Δ*BEA*, Δ*CAB*, Δ*DBC*, Δ*ECD*, Δ*ADE*, Δ*PEB*, Δ*QAC*, Δ*RBD*, Δ*SCE*, and Δ*TDA* are all isosceles.

3. Triangles Δ*DAC* and Δ*QCD* are similar (as are many others, as shown in Figure 7.21).

4. All diagonals of the pentagon are of the same length.

5. Every side of the pentagon is parallel to the diagonal "facing" it.

6. Common ratios are embedded in the figure, such as the following:
   $$\frac{AD}{DC} = \frac{CQ}{QD}.$$

7. The intersection point of two diagonals partitions both diagonals in the golden section.

8. *PQRST* is a regular pentagon.

Figure 7.21 shows how the pentagon and the pentagram relate to one another and practically infinitely approach a point.

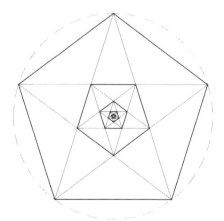

Figure 7.21.

CONSTRUCTING A REGULAR HEXAGON

We begin constructing a hexagon by drawing a circle. Then, from any point on the circle, we draw a circle of equal radius. We then continue that process, each time placing the center at the point where the previous circle intersects the original circle. You will always end up at the starting point, and you will have constructed a regular hexagon. A lot of the beautiful symmetry in Figure 7.22 is left to the reader to discover.

Figure 7.22.

CONSTRUCTING A REGULAR PENTAGON

The construction of a regular pentagon is more complicated than that for most other regular polygons. Were we to try to construct a regular pentagon in a way similar to that for other polygons, we would find ourselves in a dilemma. Consider the following curious situation.

Perhaps the most important artist that Germany has contributed to western culture is Albrecht Dürer (1471–1528). One largely forgotten work of his, which he produced in 1525, is a geometric construction (using only straightedge and compasses) of a regular pentagon. He knew that it was only an approximation of a regular pentagon, but it is extremely close to perfect, so much so that its inaccuracy is not visually detectable. Dürer offered this construction to the mathematical community as an easy alternative method to draw a regular pentagon despite the fact that the resulting shape was off by about half a degree (see C. J. Scriba and P. Schreiber, *5000 Jahre Geometrie: Geschichte,*

Kulturen, Menschen [Berlin, Germany: Springer, 2000], 259, 289–290). Although its deviation from a perfect regular pentagon is minuscule, the discrepancy cannot be ignored. Until recently, engineering books still presented Dürer's method for constructing a regular pentagon. We shall show it here, despite its flaws, since it is instructive and was seriously used for many years.

We begin with a segment, *AB* (Figure 7.23). Five circles of radius *AB* are constructed as follows:

1. Circles with centers at *A* and *B* are drawn and intersect at *Q* and *N*.
2. Draw a circle with center *Q* to intersect circles *A* and *B* at points *R* and *S*, respectively.
3. *QN* intersects circle *Q* at *P*.
4. *SP* and *RP* intersect circles *A* and *B* at points *E* and *C*, respectively.
5. Draw circles with centers at *E* and *C*, with radius *AB* to intersect at *D*.

Polygon *ABCDE* is (approximately) a regular pentagon.

Although the pentagon looks regular, the measure of angle *ABC* is about $\frac{22}{60} = \frac{11}{30}$ of a degree too large. For *ABCDE* to be a regular pentagon, each angle would have to be 108°. We will show the curious reader here that $\angle ABC \approx 108.3661202°$.

In rhombus *ABQR*, shown in Figure 7.24, $\angle ARQ = 60°$ and $BR = AB\sqrt{3}$, since *BR* is actually twice the length of an altitude of equilateral triangle *ARQ*. Since triangle *PRQ* is an isosceles right triangle, $\angle PRQ = 45°$ and $\angle BRC = 15°$.

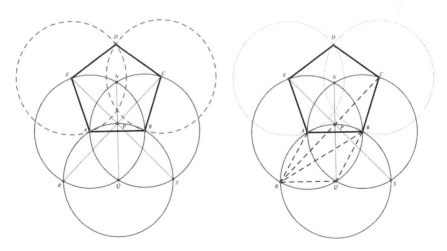

Figure 7.23.   **Figure 7.24.**

We shall apply the law of sines to $\triangle BCR$: $\dfrac{BR}{\sin\angle BCR} = \dfrac{BC}{\sin\angle BRC}$.

That is, $\dfrac{AB\sqrt{3}}{\sin\angle BCR} = \dfrac{AB}{\sin 15°}$ or $\sin\angle BCR = \sqrt{3}\sin 15°$. Therefore, $\angle BCR \approx$ 26.63387984.

In triangle BCR, $\angle RBC = 180° - \angle BRC - \angle BCR \approx 180° - 15° -$ 26.63387984° $\approx 138.3661202°$.

Thus, since $\angle ABR = 30°$, $\angle ABC = \angle RBC - \angle ABR \approx 138.661202°$ $-30° \approx 108.3661202°$, and *not* 108° as it should be in order for the shape to be a regular pentagon. Furthermore, consider the results of Dürer's construction: $\angle ABC = \angle BAE \approx 108.37°$, $\angle BCD = \angle AED \approx 107.94°$, and $\angle EDC \approx$ 107.38°.

One way to construct a proper regular pentagon would be to first construct a golden triangle and then simply mark off its base length along a given circle, as shown in Figure 7.25.

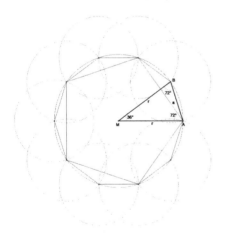

Figure 7.25.

$\mathcal{8}$

Geometric Mistakes

Up to this point we have experienced some of the beauty and amazing relationships that geometry offers us. Now it might be instructive to see how geometry can also be deceiving. Some geometric pictures are grossly misleading, and others are logically wrong. In this chapter we will entertain ourselves with these deceptions.

Depictions of geometric figures can be deceiving in a number of ways. For example, we can make mistakes in our optical perceptions. Geometry is often referred to as the visual part of mathematics, and we tend to believe many things as we see them. Consequently, geometric diagrams still play an important role in determining geometric properties and proving geometric relationships. The importance of geometric diagrams should not be minimized; however, they should be carefully analyzed, as we will see throughout this chapter. Although geometric proofs can be done without seeing a diagram, picturing the geometric figures can be very helpful. But they can still be deceiving. (For examples of mistakes in mathematics, see A. S. Posamentier and I. Lehmann, *Magnificent Mistakes in Mathematics* [Amherst, New York: Prometheus Books, 2013].)

As mentioned above, we can easily make mistakes in our visual assessment of a geometric figure. We now present some of these optical mistakes, as studying them can help make you more discriminating with visual presentations. We will first show some of these erroneous visual assessments. Then we will show how logical mistakes can be compounded and overlooked. So, follow along as we explore some of the counterintuitive characteristics that can lead to geometric mistakes!

OPTICAL MISTAKES

We begin by comparing the two segments in Figure 8.1. The one on the right side looks longer. In Figure 8.2 the bottom segment looks longer. In reality, the segments have the same length.

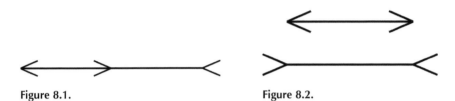

Figure 8.1.   **Figure 8.2.**

In Figure 8.3, the crosshatched segment appears longer than the clear one. In the right side of Figure 8.4, the narrower and vertical stick appears to be longer than the other two, even though to the left they are shown to be the same length.

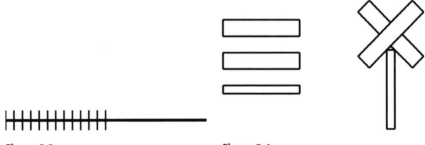

Figure 8.3.   **Figure 8.4.**

A further optical illusion can be seen in Figure 8.5, where AB appears to be longer than BC. This is not true, since $AB = BC$.

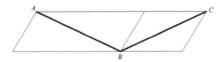

Figure 8.5.

In Figure 8.6 the vertical segment clearly appears longer, but it isn't. The curve lengths and curvature of the diagrams in Figure 8.7 look quite dissimilar. Yet, the curves are congruent!

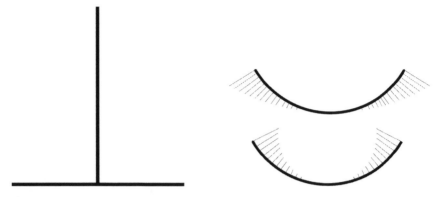

Figure 8.6.   **Figure 8.7.**

The square between the two semicircles in Figure 8.8 looks bigger than that to the left, but the two squares are the same size. In Figure 8.9 the square within the large black square looks smaller than that to the right; but, again, that is an optical illusion, since they are the same size.

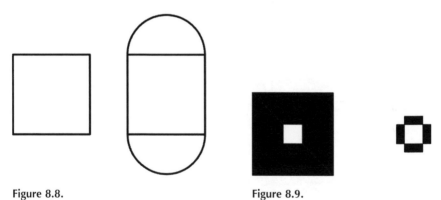

Figure 8.8.   **Figure 8.9.**

The senses are again fooled in Figure 8.10. Here the larger circle inscribed in the square (on the left) appears to be smaller than the circle circumscribed about the square (on the right). Again, the circles are the same size!

Figure 8.10.

Figures 8.11, 8.12, and 8.13 show how relative placement can affect the appearance of a geometric diagram. In Figure 8.11 the center square appears to be the largest of the group, but it isn't. In Figure 8.12 the black center circle on the left appears to be smaller than the black center circle on the right, and again it is not.

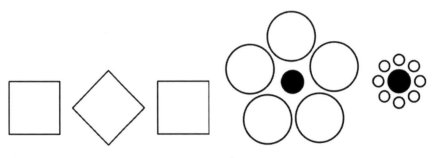

Figure 8.11.   **Figure 8.12.**

In Figure 8.13, the center sector on the left appears to be smaller than the center sector on the right. In all of these cases the two figures that appear not to be the same size are, in fact, the same size!

Figure 8.13.

Throughout this book we have avoided proving the beautiful relationships that geometry has to offer. Now, however, we will revert to "proofs" to show how geometry can also be entertaining when it claims through faulty arguments that the absurd is true. The trick is to find where the error resides. Let the reader now take on the challenge!

HOW CAN A RIGHT ANGLE EQUAL AN OBTUSE ANGLE?

This geometric mistake points out a few properties that must hold and cannot be ignored. Furthermore, it shines a spotlight on a rarely recognized concept: the *reflex angle*. Follow along as we proceed to "prove" that a right angle can be equal to an obtuse angle (an angle that is greater than 90°).

We begin with a rectangle *ABCD*, where *FA* = *BA*, *R* is the midpoint of *BC*, and *N* is the midpoint of *CF* (Figure 8.14). We will now "prove" that right angle *CDA* is equal to obtuse angle *FAD*.

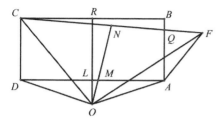

Figure 8.14.

To set up the "proof" we first draw *RL* perpendicular to *CB*, and draw *MN* perpendicular to *CF*. Then *RL* and *MN* intersect at point *O*. If they did not intersect, then *RL* and *MN* would be parallel. This would mean that *CB* is parallel to or coincides with *CF*, which is impossible. To complete the diagram for our "proof," we draw line segments *DO*, *CO*, *FO*, and *AO*.

We are now ready to embark on the "proof." Since *RO* is the perpendicular bisector of *CB* and *AD*, we know that *DO* = *AO*. Similarly, since *NO* is the perpendicular bisector of *CF*, we get *CO* = *FO*. Furthermore, since *FA* = *BA*, and *BA* = *CD*, we can conclude that *FA* = *CD*. This enables us to establish $\triangle CDO \cong \triangle FAO$ (SSS), so that $\angle ODC = \angle OAF$. We continue with *OD* = *OA*, which makes triangle *AOD* isosceles and the base angles *ODA* and *OAD* equal. Now, $\angle ODC - \angle ODA = \angle OAF - \angle OAD$ or $\angle CDA = \angle FAD$. This says that a right angle is equal to an obtuse angle. There must be some mistake!

Clearly, there is nothing wrong with this "proof." However, if you use a ruler and compasses to reconstruct the diagram, it will look like Figure 8.15.

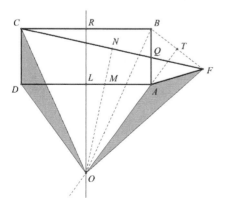

Figure 8.15.

As you see, the mistake here rests with a reflex angle—one that is often not considered. For rectangle *ABCD*, the perpendicular bisector of *AD* will also be the perpendicular bisector of *BC*. Therefore, *OC* = *OB*, *OC* = *OF*, and *OB* = *OF*. Since both points *A* and *O* are equidistant from the endpoints of *BF*, line *AO* must be the perpendicular bisector of *BF*. This is where the fault lies: we must consider the reflex angle of angle *BAO*. Although the triangles are congruent, our ability to subtract the specific angles no longer exists. Thus, the difficulty with this "proof" lies in its dependence upon an incorrectly drawn diagram.

A MISTAKEN "PROOF" THAT EVERY ANGLE IS A RIGHT ANGLE

We begin this demonstration with quadrilateral *ABCD*, where *AB* = *CD* and right angle $\angle BAD = \delta$ (see Figure 8.16). We will allow $\angle ADC = \delta'$ to be of random measure but show that it is actually a right angle. By showing this, we will have proved that any random angle is a right angle.

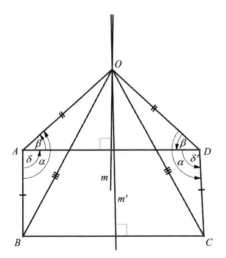

Figure 8.16.

We then construct *m*, the perpendicular bisector of *AD*, and *m'*, the perpendicular bisector of *BC*. These perpendicular bisectors intersect at point *O*. The point *O* is then equidistant from points *A* and *D*, as well as from points *B* and *C*. Therefore, *OA* = *OD* and *OB* = *OC*. We can then conclude that $\triangle OAB \cong \triangle ODC$, and it follows that $\angle BAO = \angle ODC = \alpha$.

Since triangle OAD is isosceles, it follows that $\angle DAO = \angle ODA = \beta$. Therefore, $\delta = \angle BAD = \angle BAO - \angle DAO = \alpha - \beta$, and $\delta' = \angle ADC = \angle ODC - \angle ODA = \alpha - \beta$.

It then follows that $\delta = \delta'$. However, this result is silly. There must be a mistake somewhere. Let's revisit the original diagram.

In fact, the diagram presented in Figure 8.16 tricked us since it was intentionally false. The key error is the point where the two perpendicular bisectors meet, which must be further beyond the quadrilateral than what was indicated. The correct diagram would look like that shown in Figure 8.17. We then have $\delta = \alpha - \beta$, however, $\delta' = 360° - \alpha - \beta$. This destroys the mistaken "proof." The ancient Greeks would likely have had difficulty determining the error, as the concept of "betweenness" was not addressed until the twentieth century. In other words, where does a point lie, between or not between other given points? We will encounter this issue again later.

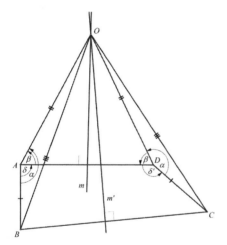

Figure 8.17.

ANOTHER MISTAKEN "PROOF" SHOWING THAT TWO
RANDOMLY DRAWN LINES IN A PLANE ARE ALWAYS PARALLEL

We begin this demonstration with the two randomly drawn lines l_1 and l_2. We then construct two parallel lines, AD and BC, that intersect our two given lines, l_1 and l_2. We complete our required diagram by drawing EF parallel to AD. The line EF intersects BD and AC in points G and H, respectively (see Figure 8.18).

The triangles AEH and ABC are similar, as are the triangles HCF and ACD. We therefore can establish the following proportions: $\dfrac{EH}{BC} = \dfrac{AH}{AC}$ and

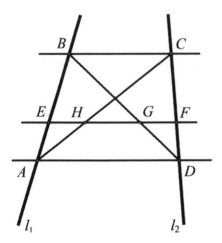

Figure 8.18.

$\dfrac{HF}{AD} = \dfrac{HC}{AC}$. When we add the two proportions we get the following:

$\dfrac{EH}{BC} + \dfrac{HF}{AD} = \dfrac{AH}{AC} + \dfrac{HC}{AC} = \dfrac{AH + HC}{AC} = \dfrac{AC}{AC} = 1$, which is to say that

$\dfrac{EH}{BC} + \dfrac{HF}{AD} = 1$.

Analogously, we can establish the similarity between triangles BGE and BDA as well as a similarity between triangles BDC and GDF and then get the following result: $\dfrac{EG}{AD} + \dfrac{GF}{BC} = 1$. Since the last two equations are equal to 1, we get

$\dfrac{EH}{BC} + \dfrac{HF}{AD} = \dfrac{EG}{AD} + \dfrac{GF}{BC}$, or $\dfrac{HF}{AD} - \dfrac{EG}{AD} = \dfrac{GF}{BC} - \dfrac{EH}{BC}$. Therefore,

$\dfrac{HF - EG}{AD} = \dfrac{GF - EH}{BC}$.

From the diagram we find that $HF - EG = (EF - EH) - (EF - GF) = GF - EH$. This tells us that the numerators of the two equal fractions are equal. Consequently, the denominators must also be equal. Therefore, $AD = BC$. Since we began with AD parallel to BC, the quadrilateral $ABCD$ must be a parallelogram, and therefore, AB is parallel to CD, or l_1 is parallel to l_2. Thus, we seem to have proved that two randomly drawn lines in the same plane are actually parallel. Clearly, this is absurd, so a mistake must have been made in this demonstration.

Let's take another look at what we have just done. From Figure 8.18 you can clearly see that $HF - EG = (HG + GF) - (EH + HG) = GF - EH$.

From the parallel lines in the diagram the following proportions follow immediately: $\dfrac{EH}{BC} = \dfrac{AE}{AB} = \dfrac{AH}{AC} = \dfrac{DF}{DC} = \dfrac{GF}{BC}$.

Since $BC \neq 0$, we then have $EH = GF$. Therefore, $GF - EH = 0$, and $HF - EG$ must also equal 0. From the earlier equation, $\dfrac{HF - EG}{AD} = \dfrac{GF - EH}{BC}$.

By substitution we have the following:

$$\frac{0}{AD} = \frac{0}{BC}.$$

This essentially tells us that we had no reason to state that $AD = BC$, since AD and BC can essentially take on any values to make this equation true. This explains where the mistake was made.

IS "PROVING" THAT A SCALENE TRIANGLE IS ISOSCELES—OR THAT ALL TRIANGLES ARE ISOSCELES—A MISTAKE?

Mistakes in geometry—also sometimes called fallacies—tend to come from faulty diagrams that result from a lack of definition. Yet, as we know, in ancient times some geometers discussed their geometric findings or relationships without a diagram. For example, as we indicated earlier, in Euclid's work the concept of "betweenness" was not considered. When this concept is omitted, we can prove that any triangle is isosceles—that is, that a triangle with three sides of different lengths actually has two sides that are equal. This sounds a bit strange. But we can demonstrate this "proof" and have the reader attempt to discover where the mistake lies before we expose it.

We shall begin by drawing a scalene triangle (i.e., a triangle with no two sides of equal length) and then "prove" it is isosceles (i.e., a triangle with two sides of equal length). Consider a scalene triangle, ABC, where we then draw the bisector of angle C and the perpendicular bisector of AB. From their point of intersection, G, we draw perpendiculars to AC and CB, meeting them at points D and F, respectively.

There are now four possibilities that match the above description for various scalene triangles: in Figure 8.19, where CG and GE meet inside the triangle at point G; in Figure 8.20, where CG and GE meet on side AB (that is, points E and G *coincide*); in Figure 8.21, where CG and GE meet outside the triangle (in G), but perpendiculars GD and GF intersect segments AC and CB (at points D and F, respectively); and in Figure 8.22, where CG and GE meet outside the triangle, but perpendiculars GD and GF intersect the extensions of sides AC and CB outside the triangle (at points D and F respectively).

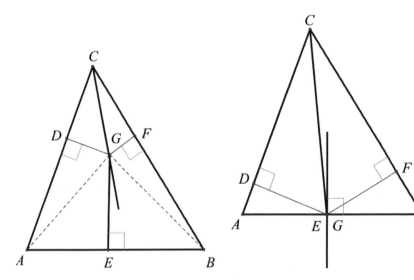

Figure 8.19.   **Figure 8.20.**

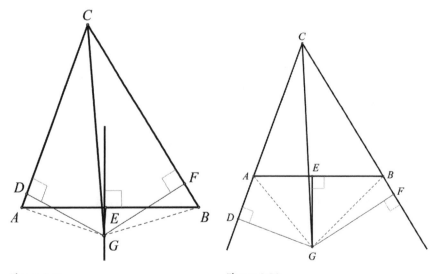

Figure 8.21.   **Figure 8.22.**

The "proof" of the mistake (or fallacy) can be done with any of the above figures. Follow along and see if the mistake shows itself without reading further. We begin with a scalene triangle, *ABC*. We will now "prove" that $AC = BC$ (or that triangle *ABC* is isosceles).

As we have an angle bisector, we have $\angle ACG \cong \angle BCG$. We also have two right angles such that $\angle CDG \cong \angle CFG$. This enables us to conclude that $\triangle CDG \cong \triangle CFG$ (SAA). Therefore, $DG = FG$ and $CD = CF$. Since a point on the perpendicular bisector (*EG*) of a line segment is equidistant from the endpoints of the line segment, $AG = BG$. Also, $\angle ADG$ and $\angle BFG$ are right angles. We then have $\triangle DAG \cong \triangle FBG$ (since they have hypotenuse and leg congruent). Therefore $DA = FB$. It then follows that $AC = BC$ (by addition in Figures 8.19, 8.20, and 8.21; and by subtraction in Figure 8.22).

At this point you may feel quite disturbed. You may wonder where the error lies that permitted this mistake to occur. You could challenge the correctness of the figures. Well, by rigorous construction you will find a subtle error in the figures. We will now divulge the mistake and show how it leads us to a better and more precise way of referring to geometric concepts.

First we can show that point *G must* be outside the triangle. Then, when perpendiculars meet the sides of the triangle, one of them will meet a side *between* the vertices, while the other will not. We can "blame" this mistake on Euclid's lack of the concept of betweenness. However, the beauty of this particular mistake lies in its proof of this betweenness issue, which establishes the mistake.

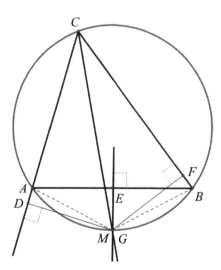

Figure 8.23.

Begin by considering the circumcircle of triangle *ABC* (Figure 8.23). The bisector of angle *ACB* must contain the midpoint, *M*, of arc *AB* (since angles *ACM* and *BCM* are congruent inscribed angles). The perpendicular bisector of *AB* must bisect arc *AB* and therefore must pass through *M*. Thus, the bisector of angle *ACB* and the perpendicular bisector of *AB* intersect *on* the circumscribed circle, which is *outside* the triangle at *M* (or *G*). This eliminates the possibilities we used in Figures 8.19 and 8.20.

Now consider the inscribed quadrilateral *ACBG*. Since the opposite angles of an inscribed (or cyclic) quadrilateral are supplementary, $\angle CAG + \angle CBG = 180°$. If angles *CAG* and *CBG* were right angles, then *CG* would be a diameter and triangle *ABC* would be isosceles. Therefore, since triangle *ABC* is scalene, angles *CAG* and *CBG* are not right angles. In this case one must be acute and the other obtuse. Suppose angle *CBG* is acute and angle *CAG* is obtuse. Then in triangle *CBG* the altitude on *CB* must be *inside* the triangle, while in obtuse triangle *CAG*, the altitude on *AC* must be *outside* the triangle. The fact that one and *only one* of the perpendiculars intersects a side of the triangle *between* the vertices destroys the fallacious "proof." This demonstration hinges on the definition of betweenness, a concept not available to Euclid.

A MISTAKEN PROOF THAT A TRIANGLE CAN HAVE TWO RIGHT ANGLES

The next geometric mistake is one that can truly upset an unsuspecting person. With two intersecting circles of any size, we draw the diameters from one of their points of intersection and then connect the other ends of the diameters, as shown in Figure 8.24.

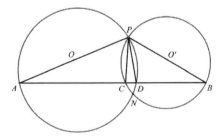

Figure 8.24.

In Figure 8.24, the endpoints of diameters *AP* and *BP* are connected by line *AB*, which intersects circle *O* at point *D* and circle *O'* at point *C*. Here, we find that $\angle ADP$ is inscribed in semicircle *PNA*, and $\angle BCP$ is inscribed in semicircle *PNB*, thus making them both right angles. We then have a dilemma: triangle *CPD* has two right angles! This is impossible. Therefore, there must be a mistake somewhere in our work.

Omission of the concept of betweenness could lead us to this dilemma. When this figure is drawn correctly, we find that angle *CPD* must equal 0, since a triangle cannot have more than 180°. That would make triangle *CPD* nonexistent. Figure 8.25 shows the correct drawing of this situation.

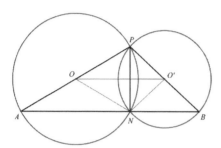

Figure 8.25.

In Figure 8.25 we can easily show that $\triangle POO' \cong \triangle NOO'$, and then $\angle POO' = \angle NOO'$. Because $\angle PON = \angle A + \angle ANO$ and $\angle ANO = \angle NOO'$ (alternate-interior angles) we have $\angle POO' = \angle A$, and then *AN* is parallel to *OO'*. The same argument can be made for circle *O'* to get *BN* parallel to *OO'*. Since line segments *AN* and *BN* are both parallel to *OO'*, they must in fact be one line, *ANB*. This proves that the diagram in Figure 8.25 is correct and the diagram in Figure 8.24 is not.

EVERY EXTERIOR ANGLE OF A TRIANGLE IS EQUAL TO ONE OF ITS REMOTE INTERIOR ANGLES

We begin with triangle *ABC*, shown in Figure 8.26, and we would like to demonstrate that angles δ and α are equal.

We now refer to Figure 8.27, where we have quadrilateral *APQC* so constructed that $\angle CAP + \angle CQP = \alpha + \varepsilon = 180°$.

We then construct a circle through three points *C*, *P*, and *Q*. Point *B* is where line *AP* intersects the circle a second time. Drawing *BC* creates a cyclic quadrilateral (i.e., one that can be inscribed in a circle), *BPQC*, where the following is true: $\angle CQP + \angle CBP = \varepsilon + \delta = \angle BCQ + \angle BPQ = 180°$.

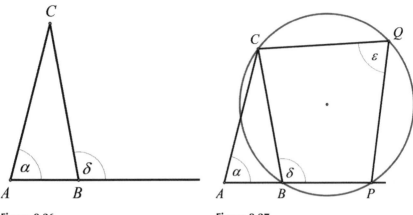

Figure 8.26.   **Figure 8.27.**

However, at the outset we had drawn $\angle CAP + \angle CQP = \alpha + \varepsilon = 180°$, so we can now conclude that $\angle CAP = \angle CBP$, which is to say that $\alpha = \delta$. Something must be wrong. Where does the mistake lie?

If quadrilateral $APQC$ has the property that $\angle CAP + \angle CQP = \alpha + \varepsilon = 180°$ and that vertices C, P, and Q lie on the same circle, then quadrilateral $APQC$ must also be cyclic, which implies that point A must also lie on the circle. This implies that points A and B must be identical. In that case, triangle ABC cannot exist. Thus, the mistake here has been revealed.

ANY POINT IN THE INTERIOR OF A CIRCLE IS ALSO ON THE CIRCLE

Let's consider the conflicting statement that any point in the interior of a circle is also on the circle. It sounds ridiculous, but we can provide a "proof" of this statement. There must be a mistake, or else we are in a logical dilemma.

We shall begin our "proof" with a circle, O, whose radius is r (see Figure 8.28). We will then let A be any point in the *interior* of the circle distinct from O, and "prove" that point A is actually *on* the circle.

We will set up our diagram as follows: let B be on the extension of OA through A such that $OA \cdot OB = OD^2 = r^2$. (Clearly OB is greater than r, since OA is less than r.) The perpendicular bisector of AB meets the circle at points D and G, where R is the midpoint of AB. We now have $OA = OR - RA$ and $OB = OR + RB = OR + RA$. Therefore, $r^2 = OA \cdot OB = (OR - RA)(OR + RA)$, or $r^2 = OR^2 - RA^2$. However, by applying the Pythagorean theorem to triangle ORD, we get $OR^2 = r^2 - DR^2$,

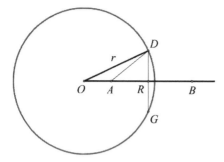

Figure 8.28.

and applying it once again to triangle *ADR* gives us $RA^2 = AD^2 - DR^2$. Therefore, since $r^2 = OR^2 - RA^2$, we get $r^2 = (r^2 - DR^2) - (AD^2 - DR^2)$, which reduces to $r^2 = r^2 - AD^2$. This would imply that $AD^2 = 0$; put another way, that *A* coincides with *D* and thus lies on the circle. That is to say, point *A* inside the circle has been proved to be on the circle. There must be a mistake somewhere!

The fallacy in this proof lies in the fact that we drew an auxiliary line *DRG* with *two* conditions—that it is the perpendicular bisector of *AB* and that it intersects the circle. Actually, all points on the perpendicular bisector of *AB* lie in the exterior of the circle, and therefore, cannot intersect the circle. Follow along with the algebraic process:

$$r^2 = OA \cdot OB$$

$$r^2 = OA(OA + AB)$$

$$r^2 = OA^2 + OA \cdot AB$$

The "proof" assumes that $OA + \dfrac{AB}{2} < r$.

By multiplying both sides of the inequality by 2 we get: $2 \cdot OA + AB < 2r$. By squaring both sides of the inequality we have: $4 \cdot OA^2 + 4 \cdot OA \cdot AB + AB^2 < 4r^2$.

By substituting four times equation (I), which is $4r^2 = 4OA^2 + 4OA \cdot AB$, into equation (II) we get $4r^2 + AB^2 < 4r^2$, or $AB^2 < 0$, which is impossible. The mistake here warns us not to allow points to take on more properties than are possible. That is, when drawing auxiliary lines, we must make sure that they use *one* condition only.

HOW CAN 64 = 65?

We now have a mathematics mistake that was popularized by Charles Lutwidge Dodgson (1832–1898), who, under the pen name of Lewis Carroll, wrote *The Adventures of Alice in Wonderland*. In Figure 8.29, we notice that the square on the left side has an area of 8 × 8 = 64 and is partitioned into two congruent trapezoids and two congruent right triangles. Yet, when these four parts are placed into a different configuration (as shown on the right side of Figure 8.29), we get a rectangle whose area is 5 × 13 = 65. How can 64 = 65? There must be a mistake somewhere.

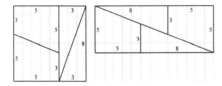

Figure 8.29.

When we correctly construct the rectangle formed by the four parts of the square, we find an extra parallelogram in the drawing—shown, exaggerated in size, in Figure 8.30.

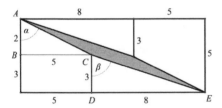

Figure 8.30.

This parallelogram (shaded) results from the fact that angles α and β are not equal. Yet, this is not easily noticeable at a glance in the original diagram! Perhaps, the easiest way to show this is to refer to the familiar tangent function. In triangle ABC, $\tan \alpha = \dfrac{5}{2} = 2.5$, while $\tan \beta = \dfrac{8}{3} \approx 2.667$. In order for line segment ACE to be a straight line—preventing a parallelogram from being formed—angles α and β would have to be equal. With different tangent values this is not the case! Thus, the mistake—one easily overlooked—has been exposed. (More such examples can be found in A. S. Posamentier and I. Lehmann, *The [Fabulous] Fibonacci Numbers* [Amherst, NY: Prometheus Books, 2007], 140–143.)

MISLEADING LIMITS

The concept of a limit is not to be taken lightly, since it is a very sophisticated one that can be easily misinterpreted. The issues surrounding the concept sometimes are quite subtle, and misunderstanding of limits can lead to some curious situations (or humorous ones, depending on your viewpoint). This point is nicely exhibited with the following two illustrations. Don't be too upset by the conclusion that you will be led to reach. Remember, this is merely for entertainment. Consider the illustrations separately and then notice their connection.

It is easy to see that the sum of the lengths of the bold segments (the "stairs") is equal to $a + b$, since the sum of the vertical bold lines equals the length $OP = a$, and the sum of the horizontal bold lines equals $OQ = b$ (see Figure 8.31).

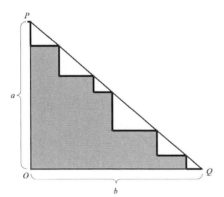

Figure 8.31.

The sum of the bold segments ("stairs"), found by adding all the horizontal and vertical segments, is $a + b$. If the number of stairs increases, the sum is still $a + b$. The dilemma arises when we continue to increase the stairs to a "limit" so that they get smaller and smaller. This makes the set of stairs appear to be a straight line, in this case the hypotenuse, PQ, of triangle POQ. It would then appear that PQ has length $a + b$. Yet we know from the Pythagorean theorem that $PQ = \sqrt{a^2 + b^2}$ and *not* $a + b$. So what's wrong?

Nothing is wrong! While the set consisting of the stairs does indeed get closer and closer to the straight line segment PQ, it does *not*, therefore, follow that the *sum* of the bold (horizontal and vertical) lengths approaches the length of PQ, contrary to our intuition. There is no contradiction here, only a failure on the part of our intuition.

Another way to "explain" this dilemma is to argue the following. As the "stairs" get smaller, they increase in number. In the most extreme situation, we have stairs of 0 length in each dimension, used an infinite number of times. This then leads to considering $0 \cdot \infty$, which is meaningless! In truth, no matter how small the stairs get, the sum of two adjacent perpendiculars that form one of the small right triangles will never be equal to their hypotenuse. They will just be small right triangles. This may be a bit difficult to see, but that is one of the dangers of working with infinity.

Just as an aside, the set of natural numbers, $\{1, 2, 3, 4, \ldots\}$, seems to be a larger set than the set of positive even numbers, $\{2, 4, 6, 8, \ldots\}$, because all the positive odd numbers are missing from the second set. Yet, since they are infinite sets, they are equal in size! We reason as follows: for every number in the set of natural numbers there is a "partner" member of the set of positive even numbers; hence they are equal in size. Counterintuitive? Yes, but that is what happens when we consider the concept of infinity.

Infinity appears to be playing games with us. The problem, however, is that with infinity we can no longer talk about the equality of sets the way we do when we have finite sets. The same is true with the staircase in our original problem. We can draw a finite number of steps, yet we cannot draw an infinite number of steps. Therein lies the problem.

A similar situation arises with the following example. In Figure 8.32 the smaller semicircles extend from one end of the large semicircle's diameter to the other.

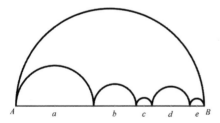

Figure 8.32.

It is easy to show that the sum of the arc lengths of the smaller semicircles is equal to the arc length of the larger semicircle; that is, the sum of the smaller semicircles

$$= \frac{\pi a}{2} + \frac{\pi b}{2} + \frac{\pi c}{2} + \frac{\pi d}{2} + \frac{\pi e}{2} = \frac{\pi}{2}(a+b+c+d+e) = \frac{\pi}{2}AB, \text{ which is the}$$

arc length of the larger semicircle. This may not appear to be true, but it is! As a matter of fact, as we increase the number of smaller semicircles (where, of course, they get smaller) their sum appears to be approaching the length of segment AB, that is, $\frac{\pi}{2} \cdot AB = AB$. Taking this a step further, if we let $AB = 1$, then we have $\pi = 2$, which we know is a mistake!

Again, the set consisting of the semicircles does indeed appear to approach the length of the straight-line segment AB. It does *not* follow, however, that the *sum* of the semicircles approaches the *length* of the limit, in this case AB.

This "apparent limit sum" is absurd, since the shortest distance between points A and B is the length of segment AB, not the semicircle arc AB (which equals the sum of the smaller semicircles). This important concept may best be explained using these motivating illustrations, so that future misinterpretations can be avoided.

OFT-MISTAKEN ATTEMPTS AT COMMON GEOMETRIC TRICKS

What is the least number of straight lines you would need to draw to connect the six points in Figure 8.33 without lifting your pencil off the paper?

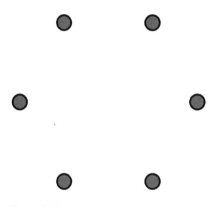

Figure 8.33.

A typical response to this question is five lines, usually drawn in one of the ways shown in Figure 8.34. But is this the *least number* of straight lines that can be used to connect these six points?

Figure 8.34.

As you might have expected, the answer is no. Fewer than five lines can be used to connect the six points. The mistake rests in the fact that we thought each line segment had to terminate at one of the points. As you can see from Figure 8.35, we were able to connect the dots with four straight lines.

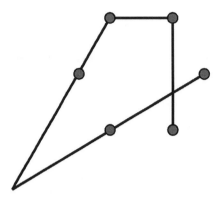

Figure 8.35.

Eliminating the restriction of having each line segment end at one of the given points allows us to get an even better solution: the three line segments shown in Figure 8.36.

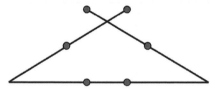

Figure 8.36.

Our earlier mistakes should now be instructive for the next situation. This time we are given nine dots, as shown in Figure 8.37, and are asked to connect them with four straight lines without lifting the pencil off the paper.

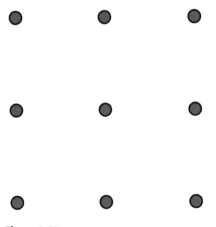

Figure 8.37.

Having learned from our earlier experiences, we should be able to arrive at the solution offered in Figure 8.38.

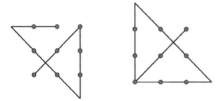

Figure 8.38.

Now that the reader will no longer make the mistake often made with the first of these dot-connecting problems, we offer two challenges. One, connect the 12 dots shown in Figure 8.39 with as few as five straight lines, without lifting the pencil off the paper, and returning to the initial point. The solution is shown in Figure 8.40.

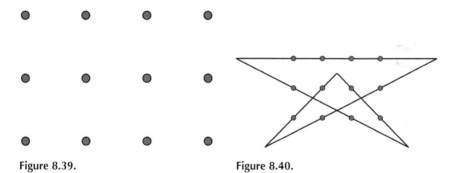

Figure 8.39.   **Figure 8.40.**

Two, connect the 25 dots shown in Figure 8.41 using only eight straight lines, without lifting the pencil off the paper, and returning to the initial point. Using nine lines would not be so difficult; using eight lines, however, is quite challenging. A solution is provided in Figure 8.42.

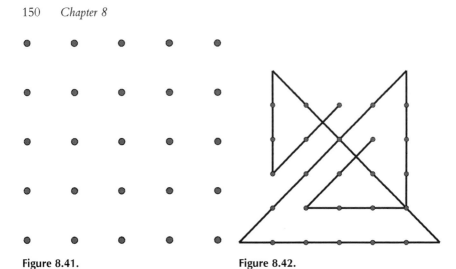

Figure 8.41.   **Figure 8.42.**

We have now seen a wide variety of geometric mistakes. Many of these give us a much stronger view of geometric principles. Those seen as "paradoxes" also allow us to see the kind of misinterpretations often encountered without notice. In sum, through exploration of geometric mistakes, our understanding of and appreciation for geometry are hugely enhanced.

References

Altshiller-Court, Nathan. *College Geometry: A Second Course in Plane Geometry for Colleges and Normal Schools*, 2nd ed. New York: Barnes and Noble, 1952.

Coxeter, H. S. M., and Greitzer, Samuel L. *Geometry Revisited.* Washington, DC: Mathematical Association of America, 1967.

Dörrie, Heinrich. *100 Great Problems of Elementary Mathematics: Their History and Solutions.* New York: Dover, 1965.

Eves, H. W. *A Survey of Geometry*, rev. ed. Boston, MA: Allyn and Bacon, 1972.

Honsberger, Ross. *Episodes in Nineteenth and Twentieth Century Euclidean Geometry.* Washington, DC: Mathematical Association of America, 1995.

Johnson, Roger A. *Modern Geometry: An Elementary Treatise on the Geometry of the Triangle and the Circle.* Boston, MA: Houghton Mifflin, 1929.

Ogilvy, C. Stanley. *Excursions in Geometry.* New York: Dover, 1990.

Pedoe, Dan. *Circles: A Mathematical View.* Washington, DC: Mathematical Association of America, 1995.

Posamentier, Alfred S. *Advanced Euclidean Geometry: Excursions for Secondary Teachers and Students.* Emeryville, CA: Key College Publishing, 2002.

———. *The Pythagorean Theorem: The Story of Its Power and Beauty.* Amherst, NY: Prometheus Books, 2010.

Posamentier, Alfred S., and Robert L. Bannister. *Geometry, Its Element and Structure.* New York: Dover Publications, 2014.

Posamentier, Alfred S., and Robert Geretschäger. *The Circle: A Mathematical Exploration Beyond the Line.* Amherst, NY: Prometheus Books, 2016.

Posamentier, Alfred S., and Ingmar Lehmann. *The Fabulous Fibonacci Numbers.* Amherst, NY: Prometheus Books, 2007.

———. *The Glorious Golden Ratio.* Amherst, NY: Prometheus Books, 2012.

———. *The Secrets of Triangles: A Mathematical Journey.* Amherst, NY: Prometheus Books, 2012.

———. *Magnificent Mistakes in Mathematics.* Amherst, NY: Prometheus Books, 2013.

————. *π: A Biography of the World's Most Mysterious Number*. Amherst, NY: Prometheus Books, 2004.

Posamentier, Alfred S., and Charles T. Salkind. *Challenging Problems in Geometry*. New York: Macmillan, 1970; reprinted by Dover Publications, 1996.

Posamentier, Alfred S., and William Wernick. *Advanced Geometric Constructions*. Dale Seymour Publications, 1988.